Proteomics Mass Spectrometry Methods

Proteomics Mass Spectrometry Methods

Sample Preparation, Protein Digestion, and Research Protocols

Edited by

Paula Meleady

*National Institute for Cellular Biotechnology,
Dublin City University, Glasnevin, Dublin, Ireland
School of Biotechnology, Dublin City University,
Glasnevin, Dublin, Ireland*

ELSEVIER

ACADEMIC PRESS
An imprint of Elsevier

Academic Press is an imprint of Elsevier
125 London Wall, London EC2Y 5AS, United Kingdom
525 B Street, Suite 1650, San Diego, CA 92101, United States
50 Hampshire Street, 5th Floor, Cambridge, MA 02139, United States

Notices

Knowledge and best practice in this field are constantly changing. As new research and experience broaden our understanding, changes in research methods, professional practices, or medical treatment may become necessary.

Practitioners and researchers must always rely on their own experience and knowledge in evaluating and using any information, methods, compounds, or experiments described herein. In using such information or methods they should be mindful of their own safety and the safety of others, including parties for whom they have a professional responsibility.

To the fullest extent of the law, neither the Publisher nor the authors, contributors, or editors, assume any liability for any injury and/or damage to persons or property as a matter of products liability, negligence or otherwise, or from any use or operation of any methods, products, instructions, or ideas contained in the material herein.

ISBN: 978-0-323-90395-0

For information on all Academic Press publications visit our website at
https://www.elsevier.com/books-and-journals

Publisher: Stacy Masucci
Acquisitions Editor: Andre Wolff
Editorial Project Manager: Makinster Barbara
Production Project Manager: Neena S. Maheen
Cover Designer: Matthew Limbert

Typeset by TNQ Technologies

Working together
to grow libraries in
developing countries

www.elsevier.com • www.bookaid.org

Contents

CHAPTER 4   Sample preparation for proteomics and mass spectrometry from clinical tissue...........55

Stephen Gargan, Paul Dowling and Kay Ohlendieck

CHAPTER 5   Sample preparation for proteomics and mass spectrometry from patient biological fluids 79

Michael Henry and Paula Meleady

CHAPTER 6   Preparation of bacterial and fungal samples for proteomic analysis..................................87

Magdalena Piatek and Kevin Kavanagh

CHAPTER 11   Profiling of the phosphoproteome using tandem mass tag labeling 163

Katie Dunphy and Paul Dowling

Contributors

Alessio Di Luca
Department of Bioscience and Agro-Food and Environmental Technology, University of Teramo, Teramo, Italy

Paul Dowling
Department of Biology, Maynooth University, National University of Ireland, Maynooth, Ireland; Kathleen Lonsdale Institute for Human Health Research, Maynooth University, National University of Ireland, Maynooth, Ireland

Katie Dunphy
Department of Biology, Maynooth University, National University of Ireland, Maynooth, Ireland; Kathleen Lonsdale Institute for Human Health Research, Maynooth University, National University of Ireland, Maynooth, Ireland

Esen Efeoglu
National Institute for Cellular Biotechnology, Dublin City University, Glasnevin, Dublin, Ireland

Stephen Gargan
Department of Biology, Maynooth University, National University of Ireland, Maynooth, Ireland; Kathleen Lonsdale Institute for Human Health Research, Maynooth University, National University of Ireland, Maynooth, Ireland

Michael Henry
National Institute for Cellular Biotechnology, Dublin City University, Glasnevin, Dublin, Ireland

Kevin Kavanagh
Department of Biology, SSPC, the Science Foundation Ireland Research Centre for Pharmaceuticals, Maynooth University, Maynooth, Co. Kildare, Ireland

Giuseppe Martino
Department of Bioscience and Agro-Food and Environmental Technology, University of Teramo, Teramo, Italy

Paula Meleady
School of Biotechnology, Dublin City University, Glasnevin, Dublin, Ireland; National Institute for Cellular Biotechnology, Dublin City University, Glasnevin, Dublin, Ireland

Luca Musante
School of Veterinary Medicine, University of Pennsylvania, Philadelphia, PA, United States

Kay Ohlendieck
Department of Biology, Maynooth University, National University of Ireland, Maynooth, Ireland; Kathleen Lonsdale Institute for Human Health Research, Maynooth University, National University of Ireland, Maynooth, Ireland

Giorgio Oliviero
Systems Biology Ireland, School of Medicine, University College Dublin, Belfield, Dublin, Ireland

Magdalena Piatek
Department of Biology, SSPC, the Science Foundation Ireland Research Centre for Pharmaceuticals, Maynooth University, Maynooth, Co. Kildare, Ireland

Karuppuchamy Selvaprakash
National Institute for Cellular Biotechnology, Dublin City University, Glasnevin, Dublin, Ireland

Kieran Wynne
Systems Biology Ireland, School of Medicine, University College Dublin, Belfield, Dublin, Ireland

Introduction

Book introduction

Paula Meleady

School of Biotechnology, Dublin City University, Glasnevin, Dublin, Ireland

Over the last decade, mass spectrometry (MS)—based proteomics has gained increased attention due to advances in sample preparation methods and instrumentation as well as improved protein databases. These advancements have led to the analysis of thousands of proteins in a sample and have made it possible to study global protein dynamics from cells to tissues and organisms. Proteins are the primary functional molecules in our cells and are responsible for almost all biological and cellular processes within the cell. In addition, the relationship between messenger RNA and protein expression is complex and not always correlated with each other for most genes and can vary significantly from protein to protein [1—4]. Proteins are also subject to a diverse array of consequential posttranslational modifications (PTMs) (e.g., phosphorylation, methylation, glycosylation, ubiquitination, etc.) which are not encoded by mRNA sequence or captured by transcript abundance. Therefore, understanding how protein abundances and their PTMs vary between healthy and disease states can provide insights into how biological activities are altered in disease conditions, thus potentially identifying new biomarkers of disease and novel therapeutic targets to improve patient outcomes.

The main idea of this methods book is to create a resource or starter guide for the non-expert proteomics scientist, as they apply proteomics techniques to answer a biological question. This book contains a range of methods' chapters from various experts in sample preparation for proteomics analysis. We have also shared some of the day-to-day protocols we use in our Core Proteomics Lab in Dublin City University, Dublin, Ireland. These methods can be used as a starting point for a user's own proteomics projects and can be optimized or modified to better suit the application being studied.

This book is divided into four main sections. Section I (Chapters 1 and 2) are introductory chapters to the book. Chapter 2 is a broad overview of proteomics and is a useful starting guide to understand the various aspects to a proteomics experiment from the initial sample preparation to analysis by mass spectrometry. Section II (Chapters 3—8) covers a range of protocols for the extraction of proteins from various biological samples, followed by methods for proteolytic digestion to peptides for mass spectrometry analysis. Chapter 3 describes a protocol for the extraction of membrane and membrane-associated proteins from mammalian cell lines and can be applied to both adherent and suspension cultured cells. Proteolytic digestion

Proteomics Mass Spectrometry Methods. https://doi.org/10.1016/B978-0-323-90395-0.00015-2

of proteins to peptides prior to LC-MS/MS analysis is described using two approaches, the filter-aided sample preparation method [5] and using a simplified kit-based approach. Chapter 4 describes a comprehensive method for the extraction of proteins from tissue samples, specifically from human muscle biopsy samples. This chapter also includes extensive guidelines on the removal of interfering chemicals from the peptide fractions prior to LC-MS/MS analysis. Chapter 5 describes a straightforward methodology for the preparation of a serum sample from a patient's blood. The method includes the use of an immunodepletion approach to remove high abundant proteins (e.g., IgG and serum albumin) from the serum sample to improve the detection of lower abundant proteins by LC-MS/MS. Chapter 6 describes methodologies for the preparation of both fungal and bacterial samples for proteomic analysis. Specifically, methods are described for sample preparation from the bacteria, *Pseudomonas aeruginosa* and *Staphylococcus aureus*, and from the fungi, *Candida albicans* and *Aspergillus fumigatus*, all of which are significant pathogens of humans. Chapter 7 describes a comprehensive method for the extraction and preparation of extracellular vesicles from urine for protein identification. Extracellular vesicles are challenging to extract from biological fluids [6]; however, have they have attracted a lot of interest in the clinical setting due to their potential to provide diagnostic information from liquid biopsies (e.g., urine, blood) in many disease indications including cancer, cardiovascular disease, and kidney disease [7−10]. Chapter 8 describes methods for protein extraction from animal samples, including information on the collection of exudates from postmortem muscle tissue.

Section III contains a protocol (Chapter 9) for in-gel digestion of proteins into peptides, with some modifications from the original published protocol [11]. This protocol is extremely useful for the extraction of protein samples from SDS-PAGE (sodium dodecyl-sulfate polyacrylamide gel electrophoresis) gels, followed by proteolytic digestion in-gel to produce peptides for mass spectrometry analysis. The method is used extensively to identify and quantify the components of protein complexes fractionated by native PAGE and in immunoprecipitation experiments using SDS-PAGE.

The final section, Section IV, includes a number of protocols and approaches to analyze the extracted peptide samples for protein identification and to identify differentially expressed proteins between experimental groups within the proteomic study being evaluated. Chapters 10 and 11 include informative protocols to identify differentially expressed proteins using both label-free and labeled (i.e., tandem mass tags) LC-MS/MS proteomic profiling and include advantages and disadvantages to both approaches. This book also includes protocols for the enrichment of common PTMs such as phosphorylation (Chapter 12) and ubiquitination (Chapter 13). The methods described in both chapters allow the site of modification to be characterized and identified, and the methods used are also amenable to differential phosphoproteomic and differential ubiquitinated proteomic studies to be carried out between experimental groups. Finally, Chapter 14 describes a high-throughput proteomics platform with the ability to analyze proteins rapidly and accurately for deep proteome coverage. It consists of an automated sample preparation press combined with an automated liquid chromatographic system, coupled with high-mass-accuracy mass spectrometry. Chapter 14 also includes a data-independent acquisition strategy for protein quantitation analysis, while Chapter 10 uses a data-dependent acquisition strategy for protein quantitation.

We are in very exciting times for the application of proteomics to understand many diseases and their progression, particularly with the development and recent launch (June 2023) of next generation high-performance mass spectrometers such as the Orbitrap Astral [12] and the TimsTOF Ultra (Bruker).[1] These instruments are capable of faster throughput, deeper coverage, and higher sensitivity with accurate and precise quantitation, compared to existing instruments on the market. These advances in MS instrumentation functionalities and sensitivity have the potential to transform our understanding of disease phenotypes to identify new biomarkers and therapeutic targets of disease, and in particular diseases where there are unmet needs.

References

[1] Payne SH. The utility of protein and mRNA correlation. Trends Biochem Sci 2015;40: 1–3. https://doi.org/10.1016/j.tibs.2014.10.010.

[2] Buccitelli C, Selbach M. mRNAs, proteins and the emerging principles of gene expression control. Nat Rev Genet 2020;21:630–44. https://doi.org/10.1038/s41576-020-0258-4.

[3] Upadhya SR, Ryan CJ. Experimental reproducibility limits the correlation between mRNA and protein abundances in tumor proteomic profiles. Cell Rep Methods 2022; 2:100288. https://doi.org/10.1016/j.crmeth.2022.100288.

[4] Jiang D, Cope AL, Zhang J, Pennell M. On the decoupling of evolutionary changes in mRNA and protein levels. Mol Biol Evol 2023;40:msad169. https://doi.org/10.1093/molbev/msad169.

[5] Wiśniewski JR, Zougman A, Nagaraj N, Mann M. Universal sample preparation method for proteome analysis. Nat Methods 2009;6:359–62. https://doi.org/10.1038/nmeth.1322.

[6] Allelein S, Medina-Perez P, Lopes ALH, Rau S, Hause G, Kölsch A, et al. Potential and challenges of specifically isolating extracellular vesicles from heterogeneous populations. Sci Rep 2021;11:11585. https://doi.org/10.1038/s41598-021-91129-y.

[7] Sun IO, Lerman LO. Urinary extracellular vesicles as biomarkers of kidney disease: from diagnostics to therapeutics. Diagnostics 2020;10:311. https://doi.org/10.3390/diagnostics10050311.

[8] Martin-Ventura JL, Roncal C, Orbe J, Blanco-Colio LM. Role of extracellular vesicles as potential diagnostic and/or therapeutic biomarkers in chronic cardiovascular diseases. Front Cell Dev Biol 2022;10:813885. https://doi.org/10.3389/fcell.2022.813885.

---

[1] https://www.bruker.com/en/news-and-events/news/2023/bruker-launches-timstof-ultra-with-csi-ultra.html

[9] Jia E, Ren N, Shi X, Zhang R, Yu H, et al. Extracellular vesicle biomarkers for pancreatic cancer diagnosis: a systematic review and meta-analysis. BMC Cancer 2022;22: 573. https://doi.org/10.1186/s12885-022-09463-x.

[10] Lee Y, Ni J, Beretov J, Wasinger VC, Graham P, Li Y. Recent advances of small extracellular vesicle biomarkers in breast cancer diagnosis and prognosis. Mol Cancer 2023; 22:33. https://doi.org/10.1186/s12943-023-01741-x.

[11] Shevchenko A, Tomas H, Havli J, Olsen JV, Mann M. In-gel digestion for mass spectrometric characterization of proteins and proteomes. Nat Protoc 2006;1:2856−60. https://doi.org/10.1038/nprot.2006.468.

[12] Heil LR, Damoc E, Arrey TN, Pashkova A, Denisov E, et al. Evaluating the performance of the Astral mass analyzer for quantitative proteomics using data-independent acquisition. J Proteome Res 2023. https://doi.org/10.1021/acs.jproteome. 3c00357.

Introduction to sample preparation for proteomics and mass spectrometry

Michael Henry[1] and Paula Meleady[1,2]

[1]*National Institute for Cellular Biotechnology, Dublin City University, Glasnevin, Dublin, Ireland;*
[2]*School of Biotechnology, Dublin City University, Glasnevin, Dublin, Ireland*

1. Introduction

Proteomics is a complex field of study which involves the application of analytical chemistry, molecular biology, biochemistry, and genetics to analyzing the structure, function, and interactions of proteins of a particular cell, tissue, or organism. Proteins are the principal functional molecules within a cell and are therefore more likely to reflect the phenotype of any living material. The inherent complexity of the proteome must also take into account that proteins are subject to post-translational modifications, trafficking, changes in subcellular location, protein—protein interactions, and complex formation, all contributing to biological function [1]. The importance of studying the proteome of a biological sample is demonstrated by the fact that the analysis of mRNA expression profiles is not necessarily a direct reflection of the protein expression profile of the cell [2]. As a result, many studies have shown a lack of correlation between mRNA and protein expression levels [3—8].

The analysis of the whole protein, typically referred to as "top-down" proteomics, is the mass analysis of an intact protein or its isoforms [9,10]. For example, within the biopharmaceutical industry whole protein analysis is required to ensure biotherapeutics are structurally defined and reproduced identically between production runs [11]. "Bottom-up" proteomics involves the identification of proteins through partial characterization of their amino acid sequence following proteolytic enzyme digestion of the intact proteins prior to mass spectrometry analysis. By comparing the masses of the proteolytic peptides or their mass spectra with those predicted from a theoretically digested sequence database, peptides can be identified and multiple peptide sequences assembled into a protein identification.

Proteomics relies on three basic technological cornerstones which include (1) a method to extract and separate complex protein or peptide mixtures, (2) mass spectrometry to acquire the data necessary to identify individual proteins, and (3) bioinformatics to analyze and assemble the MS data.

Proteomics Mass Spectrometry Methods. https://doi.org/10.1016/B978-0-323-90395-0.00008-5

2. Sample preparation for proteomic analysis

Due to the complexity of the proteome, there is no one standard method for preparing protein samples for analysis by proteomics [12]. Protein extraction is the first step of sample preparation for proteomics and includes methodologies based on detergent-based lysis, organic solvents, sample disruption by sonication, freeze/thaw cycles, mechanical disruption, or combinations of these methods. Common lysis buffer components include denaturants (e.g., urea and thiourea), ionic detergents (e.g., SDS), and nonionic detergents (Triton X-100, NP-40) to lyse cells and solubilize proteins. Proteomic protocols vary depending on the sample number, type, experimental goals, and analysis method used. Many factors are considered when designing sample preparation strategies, which can include the sample source, type, physical properties, sample abundance, complexity, and cellular location of the proteins. Each technique has its own advantage and disadvantage. For example, the use of detergents is well documented for protein extraction efficiency; however, their presence down stream of sample preparation can cause issues when using liquid chromatography and with MS instruments. Sample preparation methods such as filter aided sample preparation (FASP) [13] and suspension trapping (S-trap) [14] can effectively remove the surfactant sodium dodecyl sulfate (SDS) from extracted protein samples. Alternatives to FASP and S-trap techniques when sample starting material is minimal hence requiring maximal proteome coverage and minimal sample loss, are solid-phase enhanced separation preparation (SP3) and in-Stage Tip (iST) techniques [15]. It must be noted that iST is not compatible with SDS [16]. There are also MS-compatible, commercially available detergents that are widely used, including Rapigest (Waters), ProteaseMax (Promega), and Invitrosol (Thermo), which degrade with the addition of heat or acidic pH conditions.

Workflows that incorporate optimized cellular lysis, subcellular fractionation, depletion of high-abundance proteins, or enrichment of select proteins can all contribute to the accurate identification and quantitation of protein samples. Enrichment and/or fractionation steps can be introduced at the protein and/or peptide level if sample complexity needs to be reduced or when a specific subset of proteins/peptides are of interest. In all cases, the quality and reproducibility of sample extraction and preparation can significantly impact the research results.

The most common technique to analyze proteins that have been prepared for proteomic studies by mass spectrometry is to digest the protein(s) into smaller peptides with an enzyme and then separate the peptides by reverse phase chromatography into a mass spectrometer by electrospray ionization (ESI) which allows sequence information to be obtained by carrying out tandem mass spectrometry (MS/MS). Trypsin is by far the most common enzyme used for proteolytic digestion of proteins into peptides [17,18].

2.1 **Sample fractionation**

Despite major improvements in sensitivity and speed of modern mass spectrometers and high-performance liquid chromatography systems, basic loading capacity and ion suppression can still limit the overall coverage of complex proteomic samples. Column overloading can cause poor chromatography such as peak fronting. MS ion suppression can negatively affect precision and accuracy of the instrument. To overcome the limitations in the dynamic range of proteomic technologies and to increase the signal-to-noise ratio, protein lysates can be separated into multiple fractions which are analyzed separately by LC-MS/MS to improve proteome coverage. The need to reduce the complexity or prefractionate the sample(s) being analyzed will always be beneficial to the analysis. A sufficient amount of starting material will also predetermine if prefractionation can be carried out on the sample. There is a wide variety of strategies available to the proteomic researcher for protein and/or peptide fractionation.

At the protein level, chromatographic separation methods like ion exchange (IEX), reverse phase, hydrophobic interaction, and size exclusion can be employed. Depletion and enrichment strategies can be employed to remove high-abundance proteins of no analytical interest and isolate target proteins/peptides in the sample. It is accepted that proteins in the circulatory system mirror an individual's physiology. Protein levels are generally determined using single-protein immunoassays in hospitals such as Prostate-Specific Antigen (PSA) in blood. High-throughput, quantitative analysis using mass spectrometry—based proteomics of plasma and serum from blood, other body fluids (e.g., saliva, urine, lung lavage, vitreous humor, pancreatic juice, etc.) or tissue can focus on the discovery of clinically relevant biomarkers. Improvements in MS sensitivity now has the potential for biomarker validation [19]. However, one of the main challenges which in clinical proteomics continues to be the dynamic range of the proteins concentrations in clinical samples, for example, blood which is greater than 10 orders of magnitude [20,21].

The levels of high abundance proteins range from milligrams to grammes per liter in serum and plasma, while those of low abundance proteins are generally less than 1 µg/L [20]. It has been suggested that the more interesting, possibly tumor-specific protein biomarkers in plasma/serum may be present at 2—3 orders of magnitude lower than current LC-MS methods allow [22]. Because of these issues, plasma/serum prefractionation methods play an important role in the reduction of their complexity, allowing an opportunity to explore tissue-derived proteins that leak into the circulation as low abundance serum/plasma proteins. Approaches for plasma/serum prefractionation include lectin affinity chromatography [23], immunodepletion [24], and an "equalization" approach which uses a combinatorial library of hexapeptide ligands coupled to beads (i.e., ProteoMiner technology) to dilute high-abundance proteins and concentrate low-abundance proteins [25,26]. No single approach is perfect, but instead, different approaches may be complementary to each other. Similar challenges of high abundance proteins interfering with the analysis of low abundance proteins can also be encountered when analyzing the secretome (e.g., conditioned/spent medium) from mammalian cell lines as interfering proteins can be found in supplements such as fetal bovine serum, or by recombinant proteins secreted by a host cell line. For example, Clabaut et al. described two methods to study the secretome profiles of human mesenchymal stem cell−derived adipocytes with a combination of cell washing steps and 3 kDa centrifuge filter steps [27]. The cell culture supernatant from a CHO-K1 culture was analyzed by LC-MS/MS out resulting in the identification of 3281 different host cell proteins [28]. The CHO-K1 cells were washed six times with Phosphate Buffer Saline solution and subsequently starved for 12 h with serum-free media to reduce high abundance protein interference.

2.1.1 Immunodepletion for clinical proteomics

Immunodepletion is one of the most commonly used prefractionation techniques that can be applied to serum and plasma samples for proteomic studies, and involves the removal of the most abundant proteins through an immunocapture-based technique. Immunodepletion methods generally apply beads linked to antibodies specific toward one or many high abundant proteins in the serum or plasma sample. This capture technique with a stationary phase allows the collection of depleted proteins in the flow-through separating them from the low abundant proteins. There are many commercially available immunodepletion resins and columns that can remove one or more of the highly abundant proteins from serum or plasma (or other biological fluids); examples include ProteoPrep (SIGMA/Merck), ProteoSpin (NORGEN BioTek Corp), Proteome Purify (R&D Systems/Bio-techne), High Select Depletion Spin Columns (Thermo Scientific), and Multiple Affinity Removal System (MARS) range (Agilent). Albumin and IgG are two of the most abundant proteins found in serum and plasma, and contribute to approximately 80% of the total protein concentration [29]. The removal of serum albumin alone results in 50% depletion of the total serum/plasma proteins and has been used for biomarker discovery in plasma samples [30]. The MARS column that can deplete the 14 highest abundance proteins present in human plasma samples has been used for biomarker discovery in patients following heart failure [31] and for the early diagnosis of colorectal cancer [32].

However, extensive immunodepletion and multifractionation is expensive, time consuming, requires large volume of sample, instrument time, and analysis time. An automated, highly reproducible, 3-hour proteomic workflow from blood droplet collection to results to quantitate 1000-protein plasma proteins with the objective of developing a proteome profile as a proteomic portrait of a person's health state was developed [33]. Post blood collection, the samples were centrifuged to collect plasma and a combination of two immunodepletion kits were used to remove of 20 of the highest abundance plasma proteins [33].

2.1.2 Subcellular fractionation and enrichment

Subcellular fractionation represents a powerful tool for expanding the depth of cellular proteome coverage. Subcellular fractionation allows dissection of intracellular organelles, including the isolation of multiprotein complexes from these organelles. As such, many low abundance proteins and a variety of signaling complexes can be enriched, whereas unfractionated whole-cell lysate analyses are dominated by the most abundant proteins.

The most common and efficient method for subcellular fractionation is the use of high-speed ultracentrifugation normally coupled with density gradients. Since different compartments of a cell have different sizes and densities, each compartment will sediment into a pellet under different centrifugal forces. Fractionation yields subproteomes which have minimal cross contamination and overlap thus resulting in an improved proteomic depth and reduced sample complexity. Disadvantages of fractionation include the need for larger quantities of starting material, long processing times, and expensive instrumentation (such as ultracentrifugation) required to carry out these techniques.

Subcellular enrichment is another common method of proteome separation that can be employed if a particular subset of proteins is of interest. This method is superior to fractionation in some regards due to the smaller starting material requirements, shorter processing times, and cheaper costs. Commercial kits are widely available and are applied to enrich proteins from various subcellular compartments. Examples are the isolation of integral membrane proteins and nuclear proteins. The largely proprietary formulation of the extraction reagents provided in these kit approaches means that users are often uninformed of the mechanism of extraction; however, their efficiency and compatibility outweigh this issue. Integral membrane proteins and membrane-associated proteins can be enriched using optimized detergents to separate membrane (hydrophobic) proteins from cytosolic (hydrophilic) proteins. Similar reagent-based kit methods can be used to enrich for nuclear, cytoskeletal, lysosomal, and mitochondrial proteins. For example, we have used a commercially available membrane protein enrichment kit to enrich membrane proteins from pancreatic cancer patient samples for biomarker discovery [34,35].

Localization of organelle proteins by Isotope tagging (LOPIT) [36,37] is a proteomic method to determine the subcellular localization of membrane proteins which uses a combination of biochemical fractionation using density gradient ultracentrifugation with multiplexed quantitative mass spectrometry. This technique was later redeveloped to enable subcellular localization of thousands of protein per experiment termed "hyperLOPIT" to include spatial resolution at the suborganelle and large protein complex level [38].

2.2 Proteolytic digestion for proteomic analysis

In-solution digestion is the most commonly used method to carry out proteolytic digestion of proteins into peptides for mass spectrometry analysis, with trypsin being the most commonly used enzyme in proteomic studies [18]. Trypsin cleaves

C-terminal to arginine or lysine which produces a mixture of peptides containing one basic amino acid residue per peptide [17]. Using ESI these peptides ionize to low charge states (generally 1+, 2+, 3+, and 4+). Peptides are generally separated using HPLC, ionized and activated to produce informative fragmentation patterns. Using software algorithms, the fragment ion of the peptide is deciphered and matched to its protein. Trypsin is commercially available from numerous vendors, purified from bovine or porcine pancreas, or as a heterologous recombinant preparation. It has been reported that trypsin exhibits lower cleavage efficiency toward lysine than arginine residues, and serial digestion with Lys-C and trypsin is often recommended for complex protein digestions to produce fewer missed cleavages in the peptide sequence [39].

There are many alternatives to trypsin for LC-MS/MS proteomic studies. Furthermore, trypsin fails to produce MS-identifiable peptides derived from the C terminus of proteins. C-terminal peptides may have any terminal residue at either side, hence leading to low charge and incompatibility with LC-MS/MS analysis and search engines. As a result, protein C-termini involved in protein integration in membranes, complex formation, and protein activity are often underrepresented in tryptic proteomic studies [40]. Lys-N is a protease option that cleaves N-terminal to basic amino acids to generate positively charged C-terminal peptides compatible with LC-MS/MS [41]. Lys-N is also an excellent choice for amino acid sequencing using electron transfer dissociation (ETD) fragmentation because the peptides generated either yield simple tandem mass spectra with c-type ions and/or are rich in sequence information [42]. Asp-N hydrolyses peptide bonds on the N-terminal side of aspartic and cysteic acid residues [43]. Glu-C is a serine protease cleaving at the C-terminal of aspartic acid and glutamic acid residues [44]. Arg-C cleaves on the C-terminal of arginine residues (it also includes sites next to proline) along with lysine residues [45]. Chymotrypsin is also a serine protease cleaving at the C-terminal of tyrosine, phenylalanine, and tryptophan [17]. A combination of AspN, GluC, and trypsin was found to increase sequence coverage, robustness, and reduce missing values when compared to a standalone tryptic digested sample using Data-Independent Acquisition (DIA) MS on tip-based fractionated plasma samples [46].

3. Application of mass spectrometry for the analysis of proteins and peptides

One of the most important developments in proteomics has been the development of mass spectrometry technology. Proteomic-based MS can obtain protein structural information such as protein mass, peptide masses, and amino acid sequence information. For example, "top-down" proteomics can analyze whole proteins directly and view protein modifications that can cause mass shifts. "Bottom-up" proteomics, also known as shotgun proteomics, can be used to identify proteins by searching peptide mass and sequence mass information against protein databases.

Proteomic-based mass spectrometry requires sample ionization. The biological sample must be converted into a desolvated ion. The two traditional methods are matrix-assisted laser desorption ionization (MALDI) or, more commonly, ESI. Nano ESI coupled with nano HPLC has become the standard LC-MS approach in the majority of proteomic labs. The electrospray ionization is produced by applying a strong electric field under atmospheric pressure to a liquid passing through a capillary needle. The electric field is obtained by applying a potential difference between the capillary and a counter electrode. This electric field will induce a charge accumulation at the liquid surface at the tip of the capillary which will break the surface to form highly charged droplets. The charged droplets pass through a heated capillary causing solvent evaporation due to a high electric field in the mass spectrometer. Using ESI, peptides will produce mainly single, double, or triple charged ions, while large peptide fragments and proteins will produce multiple charges ions. Once a sample is charged, it undergoes separation, deflection, and manipulation by the mass spectrometer. The charged samples are deflected at different angles based on their individual masses. Basically, as the charged particle beam passes through the magnetic field, it undergoes separation based on the m/z ratio of its particles. The separated particles arrive at different locations on a detector within the mass spectrometer, with each location translated into a molecular ion peak on the spectrum graph.

Protein identification is obtained by amino acid sequencing, known as tandem mass spectrometry (MS/MS), which is used to fragment a specific peptide or protein which can then be used to deduce the amino acid sequence. Shotgun proteomics ("Bottom up") is the term used to describe the enzymatic digestion of proteins to produce peptides which are ionized by a mass spectrometer. The ionized peptides are analyzed by their mass-to-charge (m/z) ratio and precursors (peptides). These can then be selected for fragmentation and have the m/z fragmented peptides determined.

The majority of proteomic experiments involve the "bottom-up" approach. With the sheer numbers and large dynamic range of peptide abundances involved, these proteomic approaches require ultra-fast peptide detection and fragmentation with extreme sensitivity which will reduce measurement times, increase and improve protein identified sequence coverage to allow for in-depth proteomic analysis.

There are many instrument formats available for proteomic researchers [47]. However, linear ion trap technology [48] coupled with Fourier Transform (FT) mass spectrometry became a popular platform for proteomics as it combines high-resolution capabilities of the FT along with the robustness, speed, and sensitivity of the ion trap. With FTMS, masses are represented by frequencies, and because frequencies can be measured with high accuracy, FTMS offers very high mass measurement accuracy. Mass accuracy, with regards to mass spectrometry, is the ratio of the m/z measurement error to the true m/z and is generally described in parts per million (ppm). The mass resolving power (resolution) of an instrument is its ability to distinguish two peaks with slightly different m/z values and their mass difference is generally described as full width at half maximum (FWHM). Accurate mass is an extremely powerful filter to confirm the identity of a compound and for identification of an unknown [49]. Makarov developed the Orbitrap analyzer coupled to a linear ion trap providing proteomic researchers with a somewhat affordable, very small, and extremely powerful analyzer [50]. More recently emphasis on ion mobility in the gas phase as peptides enter the MS instruments have shown an improvement in proteome coverage. Technologies such as high-field asymmetric ion mobility (FAIMS) [51] and trapped ion mobility spectrometry (TIMS) [52] can reduce the complexity of the full MS scan.

3.1 Bottom up/shotgun proteomics

All proteins of interest are extracted and digested with a protease or combination of proteases to produce peptides and subjecting them to a "shotgun" analysis, also referred to as "bottom-up" analysis. The key steps for this approach are (1) the protein sample needs to be digested into short peptides (which can be separated by liquid chromatography), and (2) when introduced into a mass spectrometer, they are fragmented to generate sequence information by matching using computational methods to a protein database to yield a protein match. The general approach is deep coverage or as much information as possible for a defined set of proteins/peptide (targeted MS). A shotgun MS proteomics approach requires no prior knowledge of the peptides present so the MS instrument can be run in a data dependent or independent mode. Peptides obtained from the enzymatic digestion are separated by LC-MS. During the analysis, eluting peptides from the chromatography are selected to a predefined criteria (e.g., peptide signal rising above the noise in a full scan mass spectrum) and fragmented, producing tandem (MS/MS) mass spectra known as data-dependent acquisition (DDA). Alternatively, all eluting peptides with a defined m/z window can be selected for fragmentation which is known as DIA.

3.1.1 Data-dependent acquisition

DDA is the most widely used approach in shotgun proteomics [53] and generally selects the most intense peptide ion from the full MS scan and selects it for fragmentation. The criteria for MS and MS^n events are defined by the user. Examples of the parameters to be selected for full MS acquisition are mass resolution, selection of monoisotopic precursors, automatic gain control target value for the ion population and full MS maximum injection time. For tandem mass spectrometry events, parameters can be mass resolution, minimal signal threshold, maximum ion injection times, number of MS^n events, and dynamic exclusion times.

One of the main advantages of ion trap mass spectrometry is the ability to carry out multiple MS/MS fragmentation steps on a single precursor molecule (e.g., a peptide) and its product ions (e.g., fragmented peptide) to create a pattern of product–precursor ion relationship to produce extensive structural information. With LC-MS ion trap experiments, the optimization of filling the ion trap and detecting the ions within the chromatographic time period of an eluting peptide is critical for successful confident identifications. These events are described as analytical cycle time and in a high-performance ion trap can be divided into four events.

1. Ion injection
2. Automatic Gain control (pre scan)
3. Isolation and activation of the peptide in the trap
4. Mass analysis by scanning ions out of the trap

Ion injection time is determined by the rate that ions enter the trap while mass analysis is a function of scan speed and its selected mass range.

With DDA methods, the time spent on optimizing/balancing MS injection times and scan rate times can drastically improve peptide identification success rates depending on the complexity (i.e., number of peptide) of the sample being analyzed. For example, short maximum ion times and fast ion scan rates (35 ms) should be used for complex peptide samples, while longer maximum ion injection times (300 ms) with slower scan rates should be used for less complex samples. For example, using an Orbitrap Fusion Tribrid Mass Spectrometer, we analyzed two samples of various concentrations and complexity and compared fast scan rates and slower scan rates. The high concentration/complex sample was a 1 µg peptide injection from a rat thyroid cell lysate sample and we compared a maximum ion time of 35 ms and a 300 ms maximum ion time. The fast scan time method resulted in 99,502 MS/MS scans, with 18,599 unique peptides resulting in the identification of 3508 rat proteins. The slower scan times resulted in 76,989 MS/MS scans and only 15,121 unique peptides resulting in only 3119 proteins being identified (see Fig. 2.1). From this data, a short max ion time is the best for complex sample analysis as more MS/MS scans were acquired resulting in more unique peptides being identified. The low concentration/complex sample was a 0.1 µg peptide fraction from a human serum neutrophil sample that had a membrane protein enrichment. We compared the maximum ion time of 35 and 300 ms maximum ion time methods. The fast scan times resulted in 38,773 MS/MS scans, but only 713 unique peptides from 333 identified human proteins. The slower scan times resulted in 22,661 MS/MS scans but had 2150 unique peptides from 686 proteins identified (Fig. 2.1). This data demonstrates that a longer max ion time is better for abundant complex sample analysis as better quality MS/MS scans using a longer ion time is potentially more important than a maximum possible MS/MS count.

A disadvantage of DDA is there can be occasional low intersample reproducibility of peptide detection as a result of random sampling which will create a "missing value" [54].

3.1.2 Data-independent acquisition

An alternative to DDA mass spectrometry is DIA. The peptide selection criteria with DDA can favor high abundant/intensity peptides. DIA methods select all peptide ions within a given mass range and above the detection limit of the LC-MS from a sample and are fragmented independently of their intensity. DIA, theoretically,

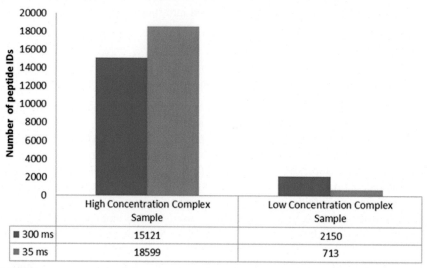

Number of Peptide IDs vs Concentration

	High Concentration Complex Sample	Low Concentration Complex Sample
■ 300 ms	15121	2150
■ 35 ms	18599	713

FIGURE 2.1

Comparison of how ion trap method parameters can drastically effect the number of confident peptide identification results. At high concentration of sample, fast ion injection time is optimal, while at low concentrations of sample, longer injection times are optimal.

allows for the identification and quantification of all peptide precursors; however, the MS/MS spectra generated is composed of the fragmented ions from all the different peptides precursors compared to DDA where the precursor peptide mass is directly linked to the MS/MS fragmentation. The acquisition workflow involves cyclical recording across the chromatographic period of consecutive full MS survey scans and MS/MS fragment ion spectra for all peptides in a predetermined isolation window. The isolation window is described as the m/z range that the MS detector uses to isolate the precursor peptide. Then, using a reference to prerecorded spectral libraries, targeted data extraction is carried out on both the full MS and MS/MS.

Peptide identification from DIA data can be achieved using reference spectral libraries for the targeted extraction of quantitative information of the peptides included in these libraries using tools such as Spectronaut, OpenSWATH, Skyline, or PeakView [55].

Another approach using DIA MS data can be a targeted approach known as Sequential Windowed Acquisition of All theoretical Fragment Ion Mass Spectra (SWATH) [55] which uses a targeted extraction of specific ion fragmented spectra for identification and quantification. SWATH MS aims to complement traditional shotgun mass spectrometry techniques and Selection Reaction Monitoring (SRM) methods [56]. An alternative approach to SWATH is WiSIM-DIA which is a combination of conventional DIA with a wide-SIM (wide selected-ion monitoring) windows which partition the precursor m/z space thus producing high quality precursor ion chromatograms [57].

3.2 MS/MS fragmentation strategies

In an observed MS/MS spectrum, the type of ions observed will depend on a number of factors including the charge state of the peptide, its primary sequence, how the energy was introduced, etc. Fragment ions can only be detected by the analyzer in a mass spectrometer if they carry at least one charge. If the retained charge is on the N-terminal fragment, the ion is classed as either a, b, or c; however, if the charge is retained on the C-terminal, then the ion is described as either x, y or z [58].

3.2.1 Collision-induced dissociation

Collision-induced dissociation (CID) is traditionally one of the most common fragmentation processes for tandem mass spectrometry to dissociate peptide ions for sequence analysis. Generally, the preferred site of cleavage of the peptide ion is its amide bond on the peptide backbone. Following CID, the amide bond on the peptide backbone will fragment to produce a series of −y ions and −b ions.

3.2.2 High collision dissociation

High collision dissociation (HCD) is a fragmentation option on Orbitrap analyzers. HCD is a higher energy dissociation compared to CID, and as a result enables a wider range of fragmentation pathways and has no low-mass cut-off [59]. Historically, HCD resulted in higher quality MS/MS spectra; however, it is a slower fragmentation strategy as spectral acquisition involves Fourier transform detection. Orbitrap mass spectrometers like the Orbitrap Fusion Tribrid instruments have the flexibility to allow for HCD fragmentation in either the Orbitrap or ion trap detector thus when detection occurs in the ion trap, it will have comparable acquisition speeds to CID ion trap detection [60].

3.2.3 Neutral loss MS3

Data dependent neutral loss (DDNL) MS3 methods consist of additional fragmentation of the product of the precursor neutral loss in the form of an MS3 scan. This technique is useful if there is a predominant loss in MS2 data that does not inform the user much about the structure of the compound (for example, in the analysis of phosphorylated peptides). To avoid this extra scan, if sufficient information is contained in the MS/MS scan, a trigger based on a dominant peak in the MS/MS spectrum can be set for a certain intensity and/or a certain mass [61]. When a phosphorylated peptide is fragmented by CID or HCD, a dominant peak corresponding to the loss of phosphoric acid (H_3PO_4), i.e., a loss of 98 Da from the parent peptide mass, is observed in the tandem mass spectrum. This loss of phosphoric acid generally results in the peptide's backbone bond staying intact, thus the resulting spectra produces a poor sequence spectrum.

3.2.4 Electron transfer dissociation

ETD is an ion/ion chemistry MS method to fragment peptides. Fragmentation occurs by transferring an electron from a radical anion to a protonated peptide. This induces fragmentation of the peptide backbone, causing cleavage of the Cα-N bond which creates complementary −c and −z type ions [62]. This is a soft/low energy fragmentation method which can preserve PTMs. ETD can also be utilized to fragment large peptides (greater than 20 amino acids) and high charge state peptide fragments which generally do not fragment well with collision dissociation fragmentation methods. Fluoranthene has emerged as one of the most widely used reagent anions for ETD [63].

3.2.5 Multistage activation

Multistage activation (MSA) is fragmentation strategy possible on hybrid mass spectrometers and more recently as CID scans on the Orbitrap Fusion Tribrid MS platforms [64]. MSA is useful when a precursor has a predominant loss that does not inform the user much about the structure of the compound which happens often in phosphorylated peptides. MSA is a pseudo-MS3 scan where the precursor is fragmented as per usual and then a subsequent specified mass below that precursor is fragmented. All of the ions are then detected in a single scan. MSA assumes that the major loss will be occurring in all cases, so the instrument does not spend time scanning out to see if that fragment is there first. MSA does have a time penalty relative to CID MS2, about 10 milli-seconds, as it is carrying out an additional activation. However, it is much quicker than performing true MS3 (neutral loss) analysis where you would first do the MS2 followed by detection then re-isolation and fragmentation for the MS3 experiment.

3.3 Identifying peptides from mass spectra

Peptide identification from mass spectra data remains one of the main bottlenecks in proteomic experiments. The first important step in proteomic data processing is the correct assignment of the MS/MS spectrum generated from a peptide ion. Identifications can be classified into two main categories:

a. Database searching
b. De novo sequencing

Database searching is the most frequently used peptide identification pipeline for large-scale proteomic studies [65]. Database searching results in a peptide identification by either correlating the MS/MS fragment ion spectra with theoretical spectra generated from a protein sequence database or by correlating the MS/MS fragment ion spectra with a spectral library of MS/MS spectra identified in previous experiments. A spectral library is meticulously complied from previously observed and identified MS/MS spectra. See Fig. 2.2 of an example using SEQUEST to match a candidate fragmented peptide to the theoretical spectrum generated from the Chinese hamster ovary FASTA protein database (UniProtKB unreviewed TrEMBL

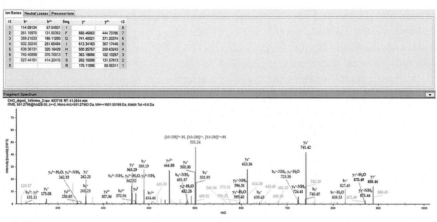

FIGURE 2.2

HCD MS/MS fragment ion spectra from a Chinese hamster ovary cell line digest analyzed on the orbitrap fusion tribrid with the data matched to the Chinese hamster ovary fasta protein database (UniProtKB unreviewed TrEMBL Chinese hamster protein database). The MS/MS spectra of b- and y-ions matched confidently with the amino acid sequence IFQIHTSR was matched to 26S protease regulatory subunit 4 [Cricetulus griseus].

Chinese hamster protein database). Theoretical b- and y—ions amino acid sequence masses (highlighted in red and blue, respectively) are confidently matched to observed spectrum masses corresponding to the amino acid sequence IFQIHTSR from the protein 26S protease regulatory subunit 4 [*Cricetulus griseus*].

De novo sequencing approach for peptide identification is a process that derives a peptide's amino acid sequence using the MS/MS fragment ion spectra without the assistance of a sequence database. It can also be used if the sequence contains modified peptides or polymorphisms. De novo sequencing is computationally heavy and requires high-resolution fragment ion spectra so generally is not compatible with large-scale proteomic studies [65].

When using protein databases for peptide identification, the MS/MS spectrum is correlated against the constructed theoretical spectra as long as a certain set of database criteria have been satisfied. One of the earliest software packages was SEQUEST [66] and then succeeded by SEQUEST HT. The main scoring function of SEQUEST is X_{corr} [66] which assigns a similarity score to a given pairing of an observed MS/MS spectrum and to that of a theoretical spectrum of a candidate peptide. The main two reasons that SEQUEST has stood the test of time is X_{corr} cross-correlation has proved to be an excellent discriminator in the presence of noise peaks and is easily integrated into fully automated processing pipelines to determine which spectra are confidently identified and then assembled protein inferences from the peptide/ion spectrum matches [67].

MASCOT software package (MATRIX SCIENCE, www.matrixscience.com) uses a probability modeling algorithm and protein database searching. Experimental peptide and fragmented ion masses are matched to those generated in silico from a database. Andromeda is a peptide search engine based on probability scoring which is integrated into the MaxQuant environment [68]. The scoring function is based on a simple binomial distribution probability formula.

Although search engines are relatively effective at identifying these peptides with a defined measure of reliability, their localization of site(s) of modification is often arbitrary and unreliable [69]. Historically, CID fragmentation is the method of choice of peptide fragmentation; however, alterative complementary configurations are now possible like ETD and HCD. Localization of modified amino acid sites using fragmentation information is not straightforward [69] as a site can only be assigned if fragment ions surrounding the site are present; this can be especially difficult if multiple sites are modified, e.g., multiple phosphorylatable residues being adjacent and single peptides having multiple modifications [70]. There are various evaluation tools available in conjunction with the database search engines to help modification site assignment including PhosRS [71], Ascore [72], and SloMo [73].

3.4 Sequence databases

The two main repositories for sequence databases are the European Bioinformatics Institute (www.ebi.ac.uk) and the National Centre of Biotechnology Information (www.ncbi.nlm.nih.gov). All mass spectrometry data search algorithms use the FASTA format and support searching of databases of protein sequences [74]. The success of the Human Genome Project making available the human genome sequence and the genomes of numerous other species has revolutionized the study of biology. Instead of investigating a single gene or a small set of genes, whole ensembles of genes can be studied simultaneously. The exact same principle allowed for the study of proteins in a comprehensive manner using mass spectrometric analysis comparing mass information against database patterns and masses.

3.5 Mass spectrometry methods for differential proteomics

Differential expression proteomics is the quantitative study of protein expression between samples that differ by some condition or variable. With this approach, protein expression of the entire proteome or of subproteomes between different samples is compared with the aim to identify novel proteins and pathways involved in specific diseases or associated with specific desirable cell phenotypes. The main approaches for differential proteomics typically involve metabolic labeling, chemical labeling or label free analysis and mass spectrometry.

3.5.1 Label-free differential expression proteomics

Mass spectrometry measurements of biological samples by direct quantitation of the signal response of their derived peptides was established over 20 years ago and demonstrated a linear relationship between protein abundance and peptide peak area [75]. Since then, numerous strategies and algorithms implementing these themes of peptide ion signal peak intensity and peptide spectral matches (PSM) spectral counting—based quantitation have been published [76].

Quantitative label-free discovery proteomics on peptide samples allows the researcher to relatively quantify and identify thousands of proteins from complex samples as long as the instrumentation can provide highly reproducible chromatography (as sample-to-sample comparison is required) and also high-resolution, accurate mass spectrometry. LC-MS runs need to be aligned for comparative expression profiling between multiple different runs/samples. There are a number of popular software packages available for label-free quantitation of peptides/proteins for MS-based proteomic experiments. Some are commercially available such as Progenesis QI for Proteomics supplied by Waters (https://www.nonlinear.com/progenesis/qi/), Proteome Discoverer MS platform from Thermo Fisher Scientific (https://www.thermofisher.com/ie/en/home/industrial/mass-spectrometry/liquid-chromatography-mass-spectrometry-lc-ms/lc-ms-software), and Scaffold from Proteome Software (https://www.proteomesoftware.com/products), while some are freely available such as MaxQuant LFQ [77]. The various software can quantify proteins across samples using the maximum (pairwise) peptide ratio information extracted peptide ion signal intensity [76] which are then normalized by minimizing the overall fold change for every peptide across all samples prior to normalization. Progenesis QI for Proteomics also quantifies each protein based on its peptide ion signal peak intensity. Recently, multiple R-packages have been evaluated for label-free quantitative proteomics experiments as they have the ability to carry out the normalization, imputation, and differential expression analysis [78].

Spectral peak intensities are the most common approach where the relative abundance of the same peptide in different samples is measured across consecutive LC-MS runs (under the assumption that the measurements are performed under identical conditions). In theory there is no limit to the number of samples that can be analyzed. Normalization is performed based on the median of the peptide ion intensities against selected reference spectra. The unique peptide ion intensity for peptides from a specific protein are summed to generate an abundance value. Using one-way ANOVA, *P*-values can be calculated based on transformed values. The supporting MS/MS data that is also generated provides peptide/protein identifications. Taking raw mass spectral data MaxQuant can assemble a 3D image from MS spectra over m/z retention times and ion intensity [79].

3.5.2 Stable isotopes tags/isobaric mass tags

Labeling approaches can be introduced at the cell growth stage using metabolic labeling or after the protein has been extracted using chemical labeling. Examples of labeling techniques include stable isotope labeling by amino acids in cell culture (SILAC), tandem mass tags (TMT), isotope-coded affinity tags (ICAT), and isobaric tags for relative and absolute quantification (iTRAQ). These techniques are all based on standard procedures for quantification by mass spectrometry in which an analyte is quantified by comparison with an introduced isotopomer of the analyte that acts as an internal standard [80]. Since the quantity of the standard is known, this means that the quantity of the analyte may be determined from the ratios of their peak intensities. Although labeling strategies can reduce MS run times, the added cost and limited number of sample labels means study design is quite fixed. Typically, you only directly compare those samples that were physically mixed and measured in one run.

Isobaric tags are very small chemical molecules with identical structure that covalently bind to the free amino termini of lysine residues of peptides thus allowing labeling peptide samples in a given experiment. Following MS/MS of labeled peptides, each isobaric tag produces a unique reporter ion signature thus making relative quantitation possible when compared to the other reporter ions intensities in the same MS/MS spectrum. The two main isobaric chemical labels used by proteomic researchers are tandem mass tag (TMT) and isobaric tag for relative and absolute quantitation (iTRAQ) [81].

Isobaric labeling via tandem mass tags enable multiplexing samples prior to LC-MS/MS allowing for high-throughput large-scale quantitative proteomics [82]. The TMT's are compounds composed of a mass reporter region, a cleavable linker region, a mass normalization region, and an amine-reactive group [80]. A major issue with using TMT can arise in the robustness at the peptide labeling stage due to the poor reaction efficiency between the peptides and the TMT labels. A recent protocol developed through the Clinical Proteomics Tumor Analysis Consortium standardized a workflow using TMT-11 across a number of proteomic laboratories [83]. The standard tandem mass tag labeling kits are 6-plex to 11-plex; however, it is now possible to label up to 18 samples with the TMTpro 18-plex label reagent set (Thermofisher.com).

3.6 Mass spectrometry for quantitative targeted proteomics

Desiderio et al. first published a targeted MS assay for peptide quantitation and used isotopically labeled peptides as internal standards [84]. Later the term Multiple Reaction Monitoring (MRM) was coined by Kusmierz et al. who compared spectra of labeled peptide samples to unlabeled in human tissue extracts [85]. Another technique, Selective Reaction Monitoring (SRM) allows the researcher to focus on the quantitation of a number of proteins of interest to perform either relative or absolute quantitation with very high specificity and sensitivity [86]. Targeted proteomics requires monitoring of proteotypic peptides (PTPs) with a unique amino acid sequence to identify a specific protein of interest in the sample proteome being analyzed by mass spectrometry [87]. The selection of the PTP is essential as targeted proteomics only involves monitoring the known precursor peptide and the fragmented ions generated from that peptide, i.e., the protein(s) under investigation are known beforehand. Using SRM it is imperative to use peptides that are specific to the candidate protein of interest and give sufficient signal intensity. In this way the experiment can be qualitatively and quantitatively validated in the samples being analyzed. Liquid chromatography separation times, ionization, and fragmentation setting must be set prior to the experiment starting. As the target peptides are known prior to analysis, MS conditions can be set up to improve selectivity and dynamic range of the instrumentation compared to conventional shotgun proteomic methods [88]. Targeted MS assays also allow for multiplexing exceeding those possible by standard immunological assays.

The classical targeted approach uses triple quadrupole mass spectrometers using the first and third quadrupoles as m/z filters and the second quadrupole as a collision cell to fragment the peptide(s) [89]. The parent and product ions are described as transitions which are further analyzed in the third quadrupole to filter specific ions thus improving selectivity and sensitivity [90]. Absolute quantitation involves spiking synthetic peptides into samples of interest at known concentrations and following mass spectrometry on the sample, the precursor and fragmented peaks of the synthetic peptide are compared to that of the natural occurring peptide in the sample [91]. Using labeled synthetic peptides, it is possible to validate biomarkers by targeted proteomics [92]. The stable isotope-labeled/stable isotope-dilution (SIL/SID) has become a gold standard for SRM protein quantitation where synthetic peptides are labeled with stable isotopes at the C-terminal arginine and lysine residues ($^{13}C_6$ $^{15}N_4$ for Arg and $^{13}C_6$ $^{15}N_2$ for Lys) that are chemically identical to the light native peptide of interest (AQUA peptide) [93]. Targeted approaches like these can be difficult to develop and maintain and this is true when methods require complex retention time scheduling which are heavily reliant on reproducible chromatography and can often require constant optimization of conditions. Newer instrumentation has allowed the move away from the traditional triple quadrupole mass spectrometers with the advent of hybrid mass spectrometer instruments including the Thermo Scientific QExactive range and Orbitrap Fusion Tribrid instruments which have the speed, selectivity, and sensitivity to allow for targeted proteomics.

Parallel reaction monitoring (PRM) is another labeled peptide targeted proteomic technique [94]. The workflow requires the addition of heavy peptides standards into the analytical sample of interest, with sequences that are analogous to endogenous peptides of interest. The heavy isotopes peptide standard differs in mass from the endogenous forms of the peptides; however, the retention times match the endogenous peptides exactly. In samples where the target peptide may be present at very low concentrations, once the specific heavy peptide is detected the instrument automatically triggers high-quality MS/MS analysis on the expected location of the endogenous form of the peptide.

Güzel et al. studying proteomic alterations in early-stage cervical cancer used PRM to quantify *MCM3*, *CEACAM5*, *S100P*, and *ICAM1* in digests of whole tissue lysates using stable isotope-labeled (SIL) peptides [95]. The urinary biomarkers MSLN and CA125 were quantified in urine from ovarian cancer patients using PRM and their abundance increased with malignancy consistent with published studies [96]. Rauniyar et al. also used PRM to quantitate multiple biomarkers in various diseases, e.g., 143 biomarkers in a kidney disease study [97]. A parallel reaction monitoring-parallel accumulation-serial fragmentation (prm-PASEF) approach was recently used to compare the absolute quantitation of proteins in human plasma using a Bruker timsTOF Pro [98]. This involved the use of synthetic peptide mixtures, including unlabeled peptides ("light" peptides—NAT) and stable-isotope—labeled standard peptides ("SIS" or "heavy" peptides), corresponding to 125 plasma proteins that were present over a broad (10^6) dynamic range [98].

4. Bioinformatic analysis of proteomic data

The output from either shotgun proteomic approaches or more targeted methods generally results in a long list of identified factors including probability score, spectral match scores, and possible quantitative values. To understand and interpret this data and to generate a hypothesis based on, for example, the response of the proteome to a challenge or condition, the protein list needs to filtered and classified. For functional analysis of a large protein list, it is important to correctly match the protein name generated from the MS search algorithm against the protein database result to a unique identifier. Generally, the first step for any functional interpretation of proteomic data is to match the protein identifier with its associated gene ontology (GO) term [99]. The three main terms that genes are associated to hierarchically clustered function terms to describe the "biological process," "molecular function," and "cellular component." Depending on the proteomic FASTA protein database used, some database search algorithms such as SEQUEST in Proteome Discoverer, MaxQuant, and X!Tandem have implemented a GO-term association step. However, if the FASTA protein database is not fully annotated with a corresponding GO term, the closest GO-term from a related protein can be located using a BLAST similarity search using the peptide identified.

Following GO-term annotation, the gene names associated with the protein list can be analyzed by GO-term enrichment analysis which will compare the abundance of the specific GO-terms with the abundance of the data list or in the organism of interest [100]. A *p*-value can be calculated to determine an overrepresentation of a specific GO-term to demonstrate functions that are significantly enriched in one sample over another. Two common web-based software tools available are DAVID and Babeleomics resources [101,102]. These tools integrate GO analysis to create and visualize protein lists in a biological context. Biological molecular pathways can also be analyzed using the gene names from the large-scale MS proteomic data generated using pathway analysis tools that are open-access databases or commercial databases. Such pathway analysis is another important feature of functional annotation. Mechanistic reasons for phenotypes can be attributed to the identification of an up or down regulated pathway from a protein list. Examples are Protein Analysis Through Evolutionary Relationships (PANTHER) classification system, Search Tool for the Retrieval of Interacting Genes/Proteins (STRING), Kyoto Encyclopedia of Genes and Genomes (KEGG), and Ingenuity Pathways Analysis (IPA) [103,104].

5. Conclusions

The application of strict sample preparation techniques, coupled with highly reproducible LC-MS/MS methods, state-of-the-art mass spectrometry, and data analytics will continue to advance the field of proteomics research generating confident reproducible data. The use of high mass accuracy, high resolution, and high sensitivity mass spectrometers will lead to in-depth identification and characterization of proteins and their modifications. Meticulous data analysis and the use of advanced data analyses tools will ensure that all proteomic outputs are of the highest quality. This will ensure proteomic research in disease indications such as cancer can produce reliable data to improve early detection, diagnosis, and response to treatment by accurately identifying changes in protein expression during disease progression, which have the potential to be novel biomarkers or new therapeutic targets of disease.

Acknowledgments

This work was supported and funded by a Science Foundation Ireland (SFI) Frontiers for the Future Award (grant no. 19/FPP/6759).

References

[1] Carbonara K, Andonovski M, Coorssen JR. Proteomes are of proteoforms: embracing the complexity. Proteomes 2021;9(3). https://doi.org/10.3390/proteomes9030038.

[2] Graves PR, Haystead TAJ. Molecular biologist's guide to proteomics. Microbiol Mol Biol Rev 2002;66:39−63. https://doi.org/10.1128/MMBR.66.1.39-63.2002.

[3] Abbott A. A post-genomic challenge: learning to read patterns of protein synthesis. Nature 1999;402:715−6. https://doi.org/10.1038/45350.

[4] Ideker T, Thorsson V, Ranish JA, Christmas R, Buhler J, et al. Integrated genomic and proteomic analyses of a systematically perturbed metabolic network. Science 2001; 292:929−34. https://doi.org/10.1126/science.292.5518.929.

[5] De Sousa Abreu R, Penalva LO, Marcotte EM, Vogel C. Global signatures of protein and mRNA expression levels. Mol Biosyst 2009;10. https://doi.org/10.1039/b908315d. 1039.b908315d.

[6] Vogel C, Marcotte EM. Insights into the regulation of protein abundance from proteomic and transcriptomic analyses. Nat Rev Genet 2012;13:227−32. https://doi.org/10.1038/nrg3185.

[7] Liu Y, Beyer A, Aebersold R. On the dependency of cellular protein levels on mRNA abundance. Cell 2016;165:535−50. https://doi.org/10.1016/j.cell.2016.03.014.

[8] Wegler C, Ölander M, Wiśniewski JR, Lundquist P, Zettl K, et al. Global variability analysis of mRNA and protein concentrations across and within human tissues. NAR Genom Bioinform 2020;2:lqz010. https://doi.org/10.1093/nargab/lqz010.

[9] Kelleher NL, Lin HY, Valaskovic GA, Aaserud DJ, Fridriksson EK, McLafferty FW. Top down versus bottom up protein characterization by tandem high-resolution mass spectrometry. J Am Chem Soc 1999;121:806—12. https://doi.org/10.1021/ja973655h.

[10] Jooß K, McGee JP, Kelleher NL. Native mass spectrometry at the convergence of structural biology and compositional proteomics. Acc Chem Res 2022;55:1928—37. https://doi.org/10.1021/acs.accounts.2c00216.

[11] Millán-Martín S, Jakes C, Carillo S, Rogers R, Ren D, Bones J. Comprehensive multi-attribute method workflow for biotherapeutic characterization and current good manufacturing practices testing. Nat Protoc 2023;18:1056—89. https://doi.org/10.1038/s41596-022-00785-5.

[12] Duong V-A, Lee H. Bottom-up proteomics: advancements in sample preparation. IJMS 2023;24:5350. https://doi.org/10.3390/ijms24065350.

[13] Manza LL, Stamer SL, Ham A-JL, Codreanu SG, Liebler DC. Sample preparation and digestion for proteomic analyses using spin filters. Proteomics 2005;5:1742—5. https://doi.org/10.1002/pmic.200401063.

[14] Zougman A, Selby PJ, Banks RE. Suspension trapping (STrap) sample preparation method for bottom-up proteomics analysis. Proteomics 2014;14:1006—10. https://doi.org/10.1002/pmic.201300553.

[15] Macklin A, Khan S, Kislinger T. Recent advances in mass spectrometry based clinical proteomics: applications to cancer research. Clin Proteonom 2020;17:17. https://doi.org/10.1186/s12014-020-09283-w.

[16] Hughes CS, Foehr S, Garfield DA, Furlong EE, Steinmetz LM, Krijgsveld J. Ultrasensitive proteome analysis using paramagnetic bead technology. Mol Syst Biol 2014;10:757. https://doi.org/10.15252/msb.20145625.

[17] Vandermarliere E, Mueller M, Martens L. Getting intimate with trypsin, the leading protease in proteomics: trypsin in proteomics. Mass Spec Rev 2013;32:453—65. https://doi.org/10.1002/mas.21376.

[18] Dau T, Bartolomucci G, Rappsilber J. Proteomics using protease alternatives to trypsin benefits from sequential digestion with trypsin. Anal Chem 2020;92:9523—7. https://doi.org/10.1021/acs.analchem.0c00478.

[19] Boys EL, Liu J, Robinson PJ, Reddel RR. Clinical applications of mass spectrometry-based proteomics in cancer: where are we? Proteomics 2023;23:2200238. https://doi.org/10.1002/pmic.202200238.

[20] Anderson NL, Anderson NG. The human plasma proteome. Mol Cell Proteomics 2002;1:845—67. https://doi.org/10.1074/mcp.R200007-MCP200.

[21] Pernemalm M, Orre LM, Lengqvist J, Wikström P, Lewensohn R, Lehtiö J. Evaluation of three principally different intact protein prefractionation methods for plasma biomarker discovery. J Proteome Res 2008;7:2712—22. https://doi.org/10.1021/pr700821k.

[22] Kim B, Araujo R, Howard M, Magni R, Liotta LA, Luchini A. Affinity enrichment for mass spectrometry: improving the yield of low abundance biomarkers. Expet Rev Proteonomics 2018;15:353—66. https://doi.org/10.1080/14789450.2018.1450631.

[23] Dahabiyeh LA, Tooth D, Barrett DA. Profiling of 54 plasma glycoproteins by label-free targeted LC-MS/MS. Anal Biochem 2019;567:72—81. https://doi.org/10.1016/j.ab.2018.12.011.

[24] Xu S, Jiang J, Zhang Y, Chen T, Zhu M, Fang C, et al. Discovery of potential plasma protein biomarkers for acute myocardial infarction via proteomics. J Thorac Dis 2019;11:3962—72. https://doi.org/10.21037/jtd.2019.08.100.

[25] Li L. Dynamic range compression with ProteoMiner™: principles and examples. In: Posch A, editor. Proteomic profiling. New York, NY: Springer New York; 2015. p. 99–107. https://doi.org/10.1007/978-1-4939-2550-6_9.

[26] Wang X, Chen M, Dai L, Tan C, Hu L, et al. Potential biomarkers for inherited thrombocytopenia 2 identified by plasma proteomics. Platelets 2022;33:443–50. https://doi.org/10.1080/09537104.2021.1937594.

[27] Clabaut A, Grare C, Léger T, Hardouin P, Broux O. Variations of secretome profiles according to conditioned medium preparation: the example of human mesenchymal stem cell-derived adipocytes: proteomics and 2DE. Electrophoresis 2015;36: 2587–93. https://doi.org/10.1002/elps.201500086.

[28] Kumar A, Baycin-Hizal D, Wolozny D, Pedersen LE, Lewis NE, et al. Elucidation of the CHO super-ome (CHO-so) by proteoinformatics. J Proteome Res 2015;14: 4687–703. https://doi.org/10.1021/acs.jproteome.5b00588.

[29] Adkins JN, Varnum SM, Auberry KJ, Moore RJ, Angell NH, et al. Toward a human blood serum proteome. Mol Cell Proteomics 2002;1:947–55. https://doi.org/10.1074/mcp.M200066-MCP200.

[30] Seong Y, Yoo YS, Akter H, Kang M-J. Sample preparation for detection of low abundance proteins in human plasma using ultra-high performance liquid chromatography coupled with highly accurate mass spectrometry. J Chromatogr B 2017;1060:272–80. https://doi.org/10.1016/j.jchromb.2017.06.023.

[31] Cao TH, Quinn PA, Sandhu JK, Voors AA, Lang CC, et al. Identification of novel biomarkers in plasma for prediction of treatment response in patients with heart failure. Lancet 2015;385:S26. https://doi.org/10.1016/S0140-6736(15)60341-5.

[32] Ahn SB, Sharma S, Mohamedali A, Mahboob S, Redmond WJ, et al. Potential early clinical stage colorectal cancer diagnosis using a proteomics blood test panel. Clin Proteonomics 2019;16:34. https://doi.org/10.1186/s12014-019-9255-z.

[33] Geyer PE, Kulak NA, Pichler G, Holdt LM, Teupser D, Mann M. Plasma proteome profiling to assess human health and disease. Cell Syst 2016;2:185–95. https://doi.org/10.1016/j.cels.2016.02.015.

[34] Coleman O, Henry M, O'Neill F, Roche S, Swan N, et al. A comparative quantitative LC-MS/MS profiling analysis of human pancreatic adenocarcinoma, adjacent-normal tissue, and patient-derived tumour xenografts. Proteomes 2018;6. https://doi.org/10.3390/proteomes6040045.

[35] Coleman O, Henry M, O'Neill F, Roche S, Swan N, et al. Proteomic analysis of cell lines and primary tumors in pancreatic cancer identifies proteins expressed only in vitro and only in vivo. Pancreas 2020;49:1109–16. https://doi.org/10.1097/MPA.0000000000001633.

[36] Dunkley TPJ, Watson R, Griffin JL, Dupree P, Lilley KS. Localization of organelle proteins by isotope tagging (LOPIT). Mol Cell Proteomics 2004;3:1128–34. https://doi.org/10.1074/mcp.T400009-MCP200.

[37] Burton JB, Carruthers NJ, Hou Z, Matherly LH, Stemmer PM. Pattern analysis of organellar maps for interpretation of proteomic data. Proteomes 2022;10:18. https://doi.org/10.3390/proteomes10020018.

[38] Mulvey CM, Breckels LM, Geladaki A, Britovšek NK, Nightingale DJ. Using hyper-LOPIT to perform high-resolution mapping of the spatial proteome. Nature Protocols 2017;12(6):1110–35. https://doi.org/10.1038/nprot.2017.026. In press.

[39] Glatter T, Ludwig C, Ahrné E, Aebersold R, Heck AJR, Schmidt A. Large-scale quantitative assessment of different in-solution protein digestion protocols reveals superior

cleavage efficiency of tandem Lys-C/trypsin proteolysis over trypsin digestion. J Proteome Res 2012;11:5145−56. https://doi.org/10.1021/pr300273g.

[40] Tsiatsiani L, Heck AJR. Proteomics beyond trypsin. FEBS J 2015;282:2612−26. https://doi.org/10.1111/febs.13287.

[41] Scholten A, Mohammed S, Low TY, Zanivan S, Van Veen TAB, Delanghe B, et al. In-depth quantitative cardiac proteomics combining electron transfer dissociation and the metalloendopeptidase Lys-N with the SILAC mouse. Mol Cell Proteomics 2011;10. https://doi.org/10.1074/mcp.O111.008474. O111.008474.

[42] Taouatas N, Drugan MM, Heck AJR, Mohammed S. Straightforward ladder sequencing of peptides using a Lys-N metalloendopeptidase. Nat Methods 2008;5: 405−7. https://doi.org/10.1038/nmeth.1204.

[43] Tetaz T, Morrison JR, Andreou J, Fidge NH. Relaxed specificity of endoproteinase Asp-N: this enzyme cleaves at peptide bonds N-terminal to glutamate as well as aspartate and cysteic acid residues. Biochem Int 1990;22:561−6.

[44] Buettner A, Maier M, Bonnington L, Bulau P, Reusch D. Multi-attribute monitoring of complex erythropoietin beta glycosylation by GluC liquid chromatography-mass spectrometry peptide mapping. Anal Chem 2020;92:7574−80. https://doi.org/10.1021/acs.analchem.0c00124.

[45] Krueger RJ, Hobbs TR, Mihal KA, Tehrani J, Zeece MG. Analysis of endoproteinase Arg C action on adrenocorticotrophic hormone by capillary electrophoresis and reversed-phase high-performance liquid chromatography. J Chromatogr 1991;543: 451−61. https://doi.org/10.1016/s0021-9673(01)95796-6.

[46] Fossati A, Richards AL, Chen K-H, Jaganath D, Cattamanchi A, Ernst JD, et al. Toward comprehensive plasma proteomics by orthogonal protease digestion. J Proteome Res 2021;20:4031−40. https://doi.org/10.1021/acs.jproteome.1c00357.

[47] Domon B, Aebersold R. Mass spectrometry and protein analysis. Science 2006;312: 212−7. https://doi.org/10.1126/science.1124619.

[48] Syka JEP, Marto JA, Bai DL, Horning S, Senko MW, et al. Novel linear quadrupole ion trap/FT mass spectrometer: performance characterization and use in the comparative analysis of histone H3 post-translational modifications. J Proteome Res 2004;3: 621−6. https://doi.org/10.1021/pr0499794.

[49] Makarov A, Denisov E, Lange O, Horning S. Dynamic range of mass accuracy in LTQ orbitrap hybrid mass spectrometer. J Am Soc Mass Spectrom 2006;17:977−82. https://doi.org/10.1016/j.jasms.2006.03.006.

[50] Hu Q, Noll RJ, Li H, Makarov A, Hardman M, Graham Cooks R. The Orbitrap: a new mass spectrometer. J Mass Spectrom 2005;40:430−43. https://doi.org/10.1002/jms.856.

[51] Cooper HJ. To what extent is FAIMS beneficial in the analysis of proteins? J Am Soc Mass Spectrom 2016;27:566−77. https://doi.org/10.1007/s13361-015-1326-4.

[52] Michelmann K, Silveira JA, Ridgeway ME, Park MA. Fundamentals of trapped ion mobility spectrometry. J Am Soc Mass Spectrom 2015;26:14−24. https://doi.org/10.1007/s13361-014-0999-4.

[53] Kalli A, Smith GT, Sweredoski MJ, Hess S. Evaluation and optimization of mass spectrometric settings during data-dependent acquisition mode: focus on LTQ-orbitrap mass analyzers. J Proteome Res 2013;12:3071−86. https://doi.org/10.1021/pr3011588.

[54] Lazar C, Gatto L, Ferro M, Bruley C, Burger T. Accounting for the multiple natures of missing values in label-free quantitative proteomics data sets to compare imputation

strategies. J Proteome Res 2016;15:1116−25. https://doi.org/10.1021/acs.jproteome.5b00981.

[55] Schubert OT, Gillet LC, Collins BC, Navarro P, Rosenberger G, et al. Building high-quality assay libraries for targeted analysis of SWATH MS data. Nat Protoc 2015;10:426−41. https://doi.org/10.1038/nprot.2015.015.

[56] Gillet LC, Navarro P, Tate S, Röst H, Selevsek N, et al. Targeted data extraction of the MS/MS spectra generated by data-independent acquisition: a new concept for consistent and accurate proteome analysis. Mol Cell Proteomics 2012;11. https://doi.org/10.1074/mcp.O111.016717. O111.016717.

[57] Koopmans F, Ho JTC, Smit AB, Li KW. Comparative analyses of data independent acquisition mass spectrometric approaches: DIA, WiSIM-DIA, and untargeted DIA. Proteomics 2018;18:1700304. https://doi.org/10.1002/pmic.201700304.

[58] Roepstorff P, Fohlman J. Letter to the editors. Biol Mass Spectrom 1984;11. https://doi.org/10.1002/bms.1200111109. 601−601.

[59] Jedrychowski MP, Huttlin EL, Haas W, Sowa ME, Rad R, Gygi SP. Evaluation of HCD- and CID-type fragmentation within their respective detection platforms for murine phosphoproteomics. Mol Cell Proteomics 2011;10. https://doi.org/10.1074/mcp.M111.009910. M111.009910.

[60] Hebert AS, Richards AL, Bailey DJ, Ulbrich A, Coughlin EE, Westphall MS, et al. The one hour yeast proteome. Mol Cell Proteomics 2014;13:339−47. https://doi.org/10.1074/mcp.M113.034769.

[61] Villén J, Beausoleil SA, Gygi SP. Evaluation of the utility of neutral-loss-dependent MS3 strategies in large-scale phosphorylation analysis. Proteomics 2008;8:4444−52. https://doi.org/10.1002/pmic.200800283.

[62] Mikesh LM, Ueberheide B, Chi A, Coon JJ, Syka JEP, Shabanowitz J, et al. The utility of ETD mass spectrometry in proteomic analysis. Biochim Biophys Acta Protein Proteonomics 2006;1764:1811−22. https://doi.org/10.1016/j.bbapap.2006.10.003.

[63] Kim M-S, Pandey A. Electron transfer dissociation mass spectrometry in proteomics. Proteomics 2012;12:530−42. https://doi.org/10.1002/pmic.201100517.

[64] Ulintz PJ, Yocum AK, Bodenmiller B, Aebersold R, Andrews PC, Nesvizhskii AI. Comparison of MS(2)-only, MSA, and MS(2)/MS(3) methodologies for phosphopeptide identification. J Proteome Res 2009;8:887−99. https://doi.org/10.1021/pr800535h.

[65] Kong AT, Leprevost FV, Avtonomov DM, Mellacheruvu D, Nesvizhskii AI. MSFragger: ultrafast and comprehensive peptide identification in mass spectrometry−based proteomics. Nat Methods 2017;14:513−20. https://doi.org/10.1038/nmeth.4256.

[66] Eng JK, Fischer B, Grossmann J, MacCoss MJ. A fast SEQUEST cross correlation algorithm. J Proteome Res 2008;7:4598−602. https://doi.org/10.1021/pr800420s.

[67] Tabb DL. The SEQUEST family tree. J Am Soc Mass Spectrom 2015;26:1814−9. https://doi.org/10.1007/s13361-015-1201-3.

[68] Cox J, Neuhauser N, Michalski A, Scheltema RA, Olsen JV, Mann M. Andromeda: a peptide search engine integrated into the MaxQuant environment. J Proteome Res 2011;10:1794−805. https://doi.org/10.1021/pr101065j.

[69] Chalkley RJ, Clauser KR. Modification site localization scoring: strategies and performance. Mol Cell Proteomics 2012;11:3−14. https://doi.org/10.1074/mcp.R111.015305.

[70] Collins MO, Wright JC, Jones M, Rayner JC, Choudhary JS. Confident and sensitive phosphoproteomics using combinations of collision induced dissociation and electron

transfer dissociation. J Proteonomics 2014;103:1—14. https://doi.org/10.1016/j.jprot.2014.03.010.

[71] Taus T, Köcher T, Pichler P, Paschke C, Schmidt A, Henrich C, et al. Universal and confident phosphorylation site localization using phosphoRS. J Proteome Res 2011; 10:5354—62. https://doi.org/10.1021/pr200611n.

[72] Beausoleil SA, Villén J, Gerber SA, Rush J, Gygi SP. A probability-based approach for high-throughput protein phosphorylation analysis and site localization. Nat Biotechnol 2006;24:1285—92. https://doi.org/10.1038/nbt1240.

[73] Bailey CM, Sweet SMM, Cunningham DL, Zeller M, Heath JK, Cooper HJ. SLoMo: automated site localization of modifications from ETD/ECD mass spectra. J Proteome Res 2009;8:1965—71. https://doi.org/10.1021/pr800917p.

[74] Cottrell JS. Protein identification using MS/MS data. J Proteonomics 2011;74: 1842—51. https://doi.org/10.1016/j.jprot.2011.05.014.

[75] Bondarenko PV, Chelius D, Shaler TA. Identification and relative quantitation of protein mixtures by enzymatic digestion followed by capillary reversed-phase liquid chromatography—Tandem mass spectrometry. Anal Chem 2002;74:4741—9. https://doi.org/10.1021/ac0256991.

[76] Al Shweiki MR, Mönchgesang S, Majovsky P, Thieme D, Trutschel D, Hoehenwarter W. Assessment of label-free quantification in discovery proteomics and impact of technological factors and natural variability of protein abundance. J Proteome Res 2017;16:1410—24. https://doi.org/10.1021/acs.jproteome.6b00645.

[77] Cox J, Hein MY, Luber CA, Paron I, Nagaraj N, Mann M. Accurate proteome-wide label-free quantification by delayed normalization and maximal peptide ratio extraction, termed MaxLFQ. Mol Cell Proteomics 2014;13:2513—26. https://doi.org/10.1074/mcp.M113.031591.

[78] Bai M, Deng J, Dai C, Pfeuffer J, Sachsenberg T, Perez-Riverol Y. LFQ-based peptide and protein intensity differential expression analysis. J Proteome Res 2023;22: 2114—23. https://doi.org/10.1021/acs.jproteome.2c00812.

[79] Schaab C, Geiger T, Stoehr G, Cox J, Mann M. Analysis of high accuracy, quantitative proteomics data in the MaxQB database. Mol Cell Proteomics 2012;11. https://doi.org/10.1074/mcp.M111.014068. M111.014068.

[80] Thompson A, Schäfer J, Kuhn K, Kienle S, Schwarz J, et al. Tandem mass tags: a novel quantification strategy for comparative analysis of complex protein mixtures by MS/MS. Anal Chem 2003;75:1895—904. https://doi.org/10.1021/ac0262560.

[81] Tian X, Permentier HP, Bischoff R. Chemical isotope labeling for quantitative proteomics. Mass Spectrom Rev 2023;42:546—76. https://doi.org/10.1002/mas.21709.

[82] Hutchinson-Bunch C, Sanford JA, Hansen JR, Gritsenko MA, Rodland KD, et al. Assessment of TMT labeling efficiency in large-scale quantitative proteomics: the critical effect of sample pH. ACS Omega 2021;6:12660—6. https://doi.org/10.1021/acsomega.1c00776.

[83] Mertins P, Tang LC, Krug K, Clark DJ, Gritsenko MA, et al. Reproducible workflow for multiplexed deep-scale proteome and phosphoproteome analysis of tumor tissues by liquid chromatography—mass spectrometry. Nat Protoc 2018;13:1632—61. https://doi.org/10.1038/s41596-018-0006-9.

[84] Desiderio DM, Kai M. Preparation of stable isotope-incorporated peptide internal standards for field desorption mass spectrometry quantification of peptides in biologic tissue. Biol Mass Spectrom 1983;10:471—9. https://doi.org/10.1002/bms.1200100806.

[85] Kusmierz JJ, Sumrada R, Desiderio DM. Fast atom bombardment mass spectrometric quantitative analysis of methionine-enkephalin in human pituitary tissues. Anal Chem 1990;62:2395−400. https://doi.org/10.1021/ac00220a026.

[86] Catenacci DVT, Liao W-L, Thyparambil S, Henderson L, Xu P, et al. Absolute quantitation of Met using mass spectrometry for clinical application: assay precision, stability, and correlation with MET gene amplification in FFPE tumor tissue. PLoS One 2014;9:e100586. https://doi.org/10.1371/journal.pone.0100586.

[87] Adeola HA, Calder B, Soares NC, Kaestner L, Blackburn JM, Zerbini LF. In silico verification and parallel reaction monitoring prevalidation of potential prostate cancer biomarkers. Future Oncol 2016;12:43−57. https://doi.org/10.2217/fon.15.296.

[88] Shi T, Song E, Nie S, Rodland KD, Liu T, Qian W, et al. Advances in targeted proteomics and applications to biomedical research. Proteomics 2016;16:2160−82. https://doi.org/10.1002/pmic.201500449.

[89] Picotti P, Clément-Ziza M, Lam H, Campbell DS, Schmidt A, et al. A complete mass-spectrometric map of the yeast proteome applied to quantitative trait analysis. Nature 2013;494:266−70. https://doi.org/10.1038/nature11835.

[90] Peterson AC, Russell JD, Bailey DJ, Westphall MS, Coon JJ. Parallel reaction monitoring for high resolution and high mass accuracy quantitative, targeted proteomics. Mol Cell Proteomics 2012;11:1475−88. https://doi.org/10.1074/mcp.O112.020131.

[91] Masuda K, Kasahara K, Narumi R, Shimojo M, Shimizu Y. Versatile and multiplexed mass spectrometry-based absolute quantification with cell-free-synthesized internal standard peptides. J Proteonomics 2022;251:104393. https://doi.org/10.1016/j.jprot.2021.104393.

[92] Gallien S, Duriez E, Domon B. Selected reaction monitoring applied to proteomics. J Mass Spectrom 2011;46:298−312. https://doi.org/10.1002/jms.1895.

[93] Gerber SA, Rush J, Stemman O, Kirschner MW, Gygi SP. Absolute quantification of proteins and phosphoproteins from cell lysates by tandem MS. Proc Natl Acad Sci USA 2003;100:6940−5. https://doi.org/10.1073/pnas.0832254100.

[94] Gallien S, Kim SY, Domon B. Large-scale targeted proteomics using internal standard triggered-parallel reaction monitoring (IS-PRM). Mol Cell Proteomics 2015;14:1630−44. https://doi.org/10.1074/mcp.O114.043968.

[95] Güzel C, Govorukhina NI, Wisman GBA, Stingl C, Dekker LJM, et al. Proteomic alterations in early stage cervical cancer. Oncotarget 2018;9:18128−47. https://doi.org/10.18632/oncotarget.24773.

[96] Sandow JJ, Rainczuk A, Infusini G, Makanji M, Bilandzic M, et al. Discovery and validation of novel protein biomarkers in ovarian cancer patient urine. Proteonomics Clin Appl 2018;12:1700135. https://doi.org/10.1002/prca.201700135.

[97] Rauniyar N, Yu X, Cantley J, Voss EZ, Belcher J, et al. Quantification of urinary protein biomarkers of autosomal dominant polycystic kidney disease by parallel reaction monitoring. Proteomics Clin Appl 2018;12:1700157. https://doi.org/10.1002/prca.201700157.

[98] Brzhozovskiy A, Kononikhin A, Bugrova AE, Kovalev GI, Schmit P-O, et al. The parallel reaction monitoring-parallel accumulation-serial fragmentation (prm-PASEF) approach for multiplexed absolute quantitation of proteins in human plasma. Anal Chem 2022;94:2016−22. https://doi.org/10.1021/acs.analchem.1c03782.

[99] Ashburner M, Ball CA, Blake JA, Botstein D, Butler H, et al. Gene ontology: tool for the unification of biology. Nat Genet 2000;25:25−9. https://doi.org/10.1038/75556.

[100] Malik R, Dulla K, Nigg EA, Körner R. From proteome lists to biological impact- tools and strategies for the analysis of large MS data sets. Proteomics 2010;10:1270–83. https://doi.org/10.1002/pmic.200900365.

[101] Medina I, Carbonell J, Pulido L, Madeira SC, Goetz S, et al. Babelomics: an integrative platform for the analysis of transcriptomics, proteomics and genomic data with advanced functional profiling. Nucleic Acids Res 2010;38:W210–3. https://doi.org/10.1093/nar/gkq388.

[102] Jiao X, Sherman BT, Huang DW, Stephens R, Baseler MW, Lane HC, et al. DAVID-WS: a stateful web service to facilitate gene/protein list analysis. Bioinformatics 2012;28:1805–6. https://doi.org/10.1093/bioinformatics/bts251.

[103] Kumar C, Mann M. Bioinformatics analysis of mass spectrometry-based proteomics data sets. FEBS (Fed Eur Biochem Soc) Lett 2009;583:1703–12. https://doi.org/10.1016/j.febslet.2009.03.035.

[104] Carrillo-Rodriguez P, Selheim F, Hernandez-Valladares M. Mass spectrometry-based proteomics workflows in cancer research: the relevance of choosing the right steps. Cancers 2023;15:555. https://doi.org/10.3390/cancers15020555.

Sample preparation

Sample preparation for proteomics and mass spectrometry from mammalian cell lines

Esen Efeoglu[1], Michael Henry[1], Paula Meleady[1,2]

[1]*National Institute for Cellular Biotechnology, Dublin City University, Glasnevin, Dublin, Ireland;*
[2]*School of Biotechnology, Dublin City University, Glasnevin, Dublin, Ireland*

1. Introduction

Proteins play a vital role in living systems by providing numerous functions such as cellular generation, replication of genetic material, acting as building blocks of cellular organelles, cell senescence, and many more [1−3]. Any disruption in the function of proteins affects the overall cellular function and may lead to several diseases [4,5]. Therefore, investigation of the cellular proteome carries undeniable importance to understanding the physiological function of proteins and the determination of possible alterations in their function in a disease state.

Proteomics is a systematic and large-scale investigation of a proteome to determine the structure, function, interaction, and expression of proteins which play role in vital processes within living systems [6]. Proteomics utilizes various technologies to investigate the proteome (Fig. 3.1). Chronologically, it is possible to say that proteomics investigation started mainly with the use of immune-specific methods (antibody-dependent) such as western blotting and ELISA, and the field of proteomics gained increased attention with nonimmune specific methods such as two-dimensional poly-acrylamide gel electrophoresis (2-D PAGE) and a version of 2-D PAGE, two-dimensional difference gel electrophoresis (2-D DIGE) [7,8]. 2-D DIGE provided simultaneous comparison and analysis of proteins from two or three experimental sample lysates on the same gel. Comparison of the proteins and/or peptides using 2D PAGE and 2D DIGE techniques required the extraction of protein lysates from biological samples using urea-based buffer. The use of another antibody-independent technique, mass spectrometry (MS), remained behind due to a lack of databases, the requirement for highly skilled operators and limited information obtained from instrumentation (low number of protein identifications) even though they provide information with high sensitivity and low detection limits (1−10 ppm).

Over the last decade, MS-based proteomics has garnered attention due to advances in sample preparation methods and instrumentation as well as improved protein databases. With the development of high-resolution mass spectrometers such as

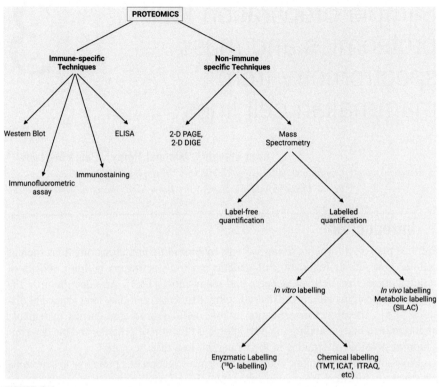

FIGURE 3.1

Various proteomic approaches. *TMT*, Tandem Mass Spectrometry; *ICAT*, Isotope-Coded Affinity Tag; *ITRAQ*, Isobaric Tags for Relative and Absolute Quantification; *SILAC*, Stable Isotope Labelling of Amino Acids in Culture.

Orbitraps, global proteomics and discovery proteomics studies significantly advanced, and the number of protein IDs increased from 100s to 1000s making global protein studies possible at cellular, tissue, and organism levels [9,10]. It is important to note that for an in-depth proteomic analysis of samples using advanced instrumentation, the quality and reliability of sample preparation remain undoubtedly important. Therefore, this methods chapter focuses on the improved sample preparation techniques from mammalian cells for MS-based proteomics, specifically focusing on more recent techniques such as filter-aided sample preparation (FASP) and in stage tip (PreOmics iST kit). In addition to these sample preparation techniques, methodologies of some upstream and downstream applications which can be applied for sample multiplexing, improved proteome coverage and MS detection are provided and discussed.

Solubilization of mammalian cells is the first step in sample preparation for proteomic analysis and this step usually requires the use of various detergents such as sodium dodecyl sulfate (SDS), CHAPS, and Triton X-100 [11,12]. Although detergents are great solubilizers of mammalian cells, they interfere with many downstream sample preparation steps for MS-based proteomics. They have been shown to suppress enzymatic digestion by altering the structure of the tryptic enzyme (denaturation), interfering with HPLC column binding and elution and contaminating the instrumentation. Therefore, in any MS-based proteomics experiment, great caution should be taken for the use of detergents to maintain the balance between effective solubilization and preventing interference of LC columns and MS instrumentation. To aid this balance, detergent removal strategies have been developed using organic compounds to precipitate the proteins and remove them from the detergent containing solution or required the use of dilutions; both of these approaches resulted in either extreme sample loss, over dilution of peptides and difficulty to scale up to obtain high amounts of peptides for analysis of posttranslational modifications which can require up to 5–20 mg of peptide [13].

The FASP technique was introduced by Manza in 2005 [14] and further modified by Wiśniewski et al. in 2009 [15], and rapidly got accepted by many proteomics research laboratories. The FASP technique offered a highly effective and clever way of detergent removal using a standard ultrafiltration device with a pore size which allows the passage of low-molecular-weight contaminating detergents but retains and concentrates the high-molecular-weight proteins. After detergent removal is complete, the ultrafiltration device further acts as a "proteomic reactor" where buffer exchange and chemical modifications can take place. Finally, digestion can be carried out in the same ultrafiltration device. The pore size of the ultrafiltration device allows for the passage of peptides and collection for MS analysis. The FASP method has been applied to a diverse range of biological samples from microorganisms to mammalian cells and tissue samples [16–18]. The technique has also been applied to cell lines which have a significant potential for biopharmaceutical production such as Chinese hamster ovary (CHO) cells [19]. In this protocol, Coleman et al. showed a step-by-step modified protocol for in-depth proteomic analysis of the recombinant CHO cell proteome using a FASP combined with strong cation exchange [19].

Following the idea of using a single "proteomic reactor" system with the introduction and improvement in FASP, in stage tip (iST) protein extraction and digestion was introduced in 2014 [20]. PreOmics GmBh introduced their MS sample preparation kits, which use the "single vessel" approach but also ease the sample preparation by combining cell lysis, processing and digestion of proteins in stage tips with C18 disks. Although the approach is the same as the in-solution digestion (cell lysis, reduction of proteins, alkylation of proteins, digestion, and clean-up), a single reactor system reduces the sample loss significantly and a combination of lysis, alkylation and digestion significantly reduces the hands-on time (from approximately 2 days to a couple of hours). The C18 disks provide a physical barrier for insoluble material and macromolecules allowing clean-up of the peptides at the end. The wash solutions provided by the kit provide not just desalting but also the removal of hydrophilic and hydrophobic contaminants. Sielaf et al. [21] have compared the FASP, iST, and single-pot solid-phase-enhanced sample preparation (SP3) techniques to observe their performance, especially in the low microgram range (1−20 µg) and showed that although all three techniques performed well at protein amounts of ∼20 µg, iST and SP3 were observed to work with higher performance with lower amounts of protein starting material. The study also showed high reproducibility and precision for iST and SP3 techniques even at protein amounts ∼1 µg.

Although these techniques can be carried out on their own and the proteome of the cells can be investigated using MS, samples can be further processed to improve MS analysis or to obtain more information from various cellular regions. Therefore, following two more recent sample preparation techniques (FASP and iST), this chapter provides protocols for membrane/cytoplasmic fractionation, fractionation of peptides, and chemical labeling. All these techniques described in this chapter show compatibility with both FASP and the PreOmics iST kit.

2. Materials and equipment

- TRIZMA HCl
- TRIZMA Base
- SDS
- DL-Dithiothreitol (DTT)
- Iodoacetamide
- Urea
- HPLC grade water
- Phosphate-buffered saline (PBS)
- Trifluoroacetic acid (TFA)
- Acetonitrile (ACN)
- Sodium chloride
- Ammonium bicarbonate
- Potassium phosphate monobasic (KH_2PO_4)
- Potassium chloride (KCl)
- Sequencing-grade modified trypsin (Promega)
- ProteaseMax Surfactant Trypsin Enhancer (Promega)
- Microcon-10kDa Centrifugal Filter Unit with Ultracel-10 membrane (Merck Millipore)
- Pierce C18 100 µL tips (Thermo Fisher Scientific)
- iST Sample preparation kit (PreOmics GmbH, Germany)
- Mem-PER Plus Membrane Protein Extraction Kit (Thermo Fisher Scientific)
- TMT10Plex Isobaric Label Reagent Set (ThermoFisher Scientific)
- Pierce BCA Protein Assay Kit (Thermo Fisher Scientific)

- Tip sonicator
- Spectrophotometer (e.g., MultiskanTM GO, Thermo Fisher Scientific)
- Vacuum evaporator
- Heating block
- Pierce Strong Cation Exchange Spin Column, Mini (Thermo Fisher Scientific)
- pH meter

3. Before you begin

3.1 Filter-aided sample preparation

The FASP method is based on the use of an ultrafiltration device which allows the passage of detergents but retains and concentrates the proteins based on the size of the membrane pores. It is a well-established and commonly used method in many proteomic facilities due to the high proteome coverage and adaptability of the technique to various mammalian cell lines. The technique requires the preparation of buffers such as Tris-HCl solution, lysis buffer, wash buffers, alkalization solution, sample acidification buffer, wetting solution, elution solution, and rinse solution.

1. *Tris-HCl solution:* Prepare 0.1 M stock solutions of Tris-HCl at pH 7.6, 7.9, and 8.5 using TRIZMA HCl, TRIZMA Base, and HPLC grade water.
2. *Lysis buffer:* Prepare 4% SDS and 0.1 M DTT solution. For example, add 0.8 g of SDS and 0.0308 g of DTT to 20 mL of 0.1 M Tris-HCl (pH 7.6). (Note 1)
3. *Wash buffers:* Prepare 8 M urea in 0.1 M Tris-HCl pH 8.5 (Wash Buffer 1). Prepare 8 M Urea in 0.1 M Tris-HCl pH 7.9 (Wash Buffer 2), e.g., 0.4 g urea in 1 mL of 0.1 M Tris-HCl.
4. *Alkylation solution:* Prepare 0.05 M Iodoacetamide in 8 M urea, 0.1 M Tris-HCl (pH 8.5). *Rinse solution:* Prepare 0.5 M NaCl in HPLC grade water.
5. *Ammonium bicarbonate solution:* Freshly prepare 50 mM ammonium bicarbonate in HPLC grade water. (Note 2)
6. *Trypsin solution:* Add 50 mM acetic acid into the lyophilized trypsin to reconstitute. The final concentration should be 1 μg/mL. (Note 3)
7. *ProteaseMax Surfactant:* Prepare 1% solution of ProteaseMax Surfactant by the addition of 100 μL of freshly prepared 50 mM ammonium bicarbonate into a 1 mg vial.
8. *Sample acidification buffer:* Prepare 2.5% TFA using HPLC grade water.
9. *Wetting solution:* Prepare 50% ACN using HPLC grade water.
10. *Solutions for equilibration and rinse of peptides:* Prepare 0.1% TFA and 2% ACN in HPLC grade water.
11. *Peptide elution solution:* Prepare 0.1% TFA and 70% ACN in HPLC grade water.

3.2 PreOmics iST kit

The PreOmics iST Sample preparation kit contains all chemicals needed for the preparation of protein digests obtained from mammalian cells. The kit includes the following: (a) a LYSE solution which denatures, reduces, and alkylates the proteins in one step, (b) RESUSPEND solution for the resuspension of DIGEST (Trypsin/Lys-C mix) delivered as a lyophilized powder, (c) DIGEST to digest proteins extracted from the mammalian cells, (d) a STOP solution to stop the enzymatic digestion, (e) WASH1 and WASH2 solutions to clean the digested peptides from hydrophobic and hydrophilic contaminants, respectively, (f) an ELUTE solution to elute peptides from the cartridge, and (g) LC-LOAD for resuspension of peptides in an LC-MS compatible solution.

In addition to the reagents included, the kit includes cartridges which allow digestion of protein samples (starting material range: 1–100 µg) (Note 4), waste and collection tubes (Note 5) and adapters for fitting cartridges into the tubes.

Sample preparation by using the PreOmics iST kit also requires equipment such as a refrigerated high-speed centrifuge (e.g., Hettich Mikro 200R), vacuum evaporator (e.g., Speed Vac), heating block, sonicator (if necessary for DNA shear), and ultrasonic bath.

3.3 Membrane fractionation

The Mem-PER Plus Membrane Protein Extraction Kit provides separation of membrane-related proteins from cytoplasmic proteins with high efficiency leading to higher proteome coverage for proteomic studies. The kit contains a cell wash solution, solubilization buffer, and permeabilization buffer. The cell wash solution is used to wash cells to remove contaminants from the cell culture medium. Permeabilization buffer (a mild detergent) allows for the release and collection of cytoplasmic proteins, whereas solubilization buffer provides solubilization of membrane proteins. The kit contains a sufficient amount of reagents for membrane fractionation of 50 mammalian cell pellets with a pellet size of 5×10^6 cells per pellet. Cell wash and solubilization solutions are stored at 4°C, whereas permeabilization buffer needs to be stored at −20°C.

In addition to the reagents/buffers provided with the kit, protease and phosphatase inhibitors are needed to reach optimal results. Equipment such as micropipettes, heating block, vortex, and centrifuge are needed to carry out the protocol.

3.4 High pH reversed-phase fractionation

The Pierce High pH reversed-phase peptide fractionation kit includes the reversed-phase fractionation spin columns (20 mg of resin in 1:1 water/DMSO slurry), 100 mL of 0.1% triethylamine prepared in water. In addition to the kit contents, TFA, ACN, and HPLC grade water are required for the fractionation.

The following solutions should be prepared prior to the experiment:

1. *0.1% TFA (equilibration) solution:* Add 10 μL of TFA into 10 mL of HPLC grade water.
2. *Wash solution and Elution solutions:* Prepare the wash solution and elution solutions according to Table 3.1.

3.5 Peptide purification and fractionation: strong cation exchange

1. *Diluent solution:* Prepare 10 mM KH_2PO_4 and 25% ACN using HPLC grade water. Adjust the pH to 3.
2. *Elution Buffers:* Prepare 10 mL of KCl elution buffers in diluent solution by adjusting the final concentration of the KCl to 10, 25, 50, 75, 100, 125, 150, 200, 300, and 500 mM (Table 3.2).

3.6 Tandem Mass Tag (TMT) labeling

TMT10Plex Isobaric Label Reagent Set includes the reagents sufficient for one 10-plex (1 × 10-way) experiments. As labeling reagents, the kit contains TMT10-126, TMT10-127N, TMT10-127C, TMT10-128N, TMT10-128C, TMT10-128N, TMT10-129C, TMT10-130N, TMT10-130C, and TMT10-131. 0.2, 0.8, or 5 mg label containing vials are available as well as smaller and larger-scale labeling kits such as TMTduplex, 6Plex TMT, 11Plex TMT, and 16Plex TMT pro. The most suitable kit can be selected based on the scale of the labeling/application.

In addition to the labels, the protocol requires the use of anhydrous ACN (LC-MS grade), LC-MS grade water, chilled (−20°C) acetone, and protein assays such as BCA. A bench-top centrifuge, a heating block, and a SpeedVac are also required as instrumentation.

Table 3.1 Preparation of elution solutions.

Fraction No	ACN (%)	ACN (μL)	0.1% Triethylamine (μL)
WASH	5	50	950
1	10	100	900
2	12.5	125	875
3	15	150	850
4	17.5	175	825
5	20	200	800
6	22.5	225	775
7	25	250	750
8	50	500	500

Table 3.2 Preparation of elution buffers for strong cation exchange.

Elution buffer	KCl concentration (mM)	KCl (mg)	Diluent solution (mL)
1	10	7	10
2	25	19	10
3	50	37	10
4	75	56	10
5	100	75	10
6	125	93	10
7	150	112	10
8	200	149	10
9	300	224	10
10	500	373	10

4. Step-by-step method details

4.1 Cell culture and collection of cell pellets for proteomics

For proteomics sample preparation, mammalian cell lines are cultured in their respective growth medium. The culturing can be carried out in cell culture dishes, flasks, or well plates depending on the application and also the required final protein concentration. If the proteomics application requires high amounts of protein, for example, analysis of the phophoproteome, cell culture plastics which provides the growth of higher density of cells should be preferred. Once cells reached 80% −90% confluence in the culture, cells are detached from the surface via trypsinization and centrifuged at 300 ×g for 5 min. For suspension cultures, cell can directly be spun down at 300 ×g for 5 min. Following centrifugation, the supernatant is removed and cells are washed with sterile PBS three times. Following the third PBS wash, PBS is removed without dispersing the cell pellet and the cell pellet can be snap frozen using liquid nitrogen for storage at −80°C. Alternatively, cell pellets can be freshly used for the proteomic analysis.

4.2 Filter-aided sample preparation

The methodology is described in detail in this section and a schematic of the workflow for the use of FASP for mass spectrometry sample preparation from mammalian cell lines is provided in Fig. 3.2.

1. Resuspend the cell pellets (including approximately 1 mg of starting protein material) in 1 mL of lysis buffer (Section 3.1.1) using a pipette (up and down).

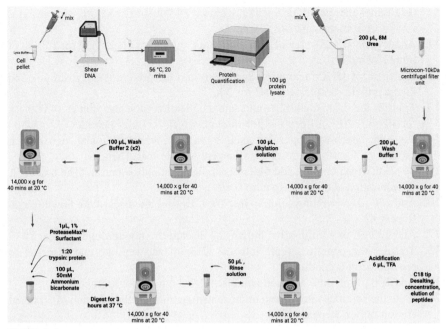

FIGURE 3.2

Workflow for filter-aided sample preparation (FASP) from a mammalian cell pellet.

2. Further disrupt the cell integrity and shear the DNA using a tip sonicator. (Note 6)

3. By using a heating block, heat the cell lysate at 56°C for 20 min to denature the proteins.

4. Use the BCA assay to determine protein concentration as per manufacturer's instructions and aliquot the protein samples (100 µg). Aliquoted samples can be kept at −80°C for further use.

5. Transfer 100 µg of protein lysate into a clean microcentrifuge tube, and add 200 µL of 8 M urea and mix well by using a micropipette.

6. Transfer the lysate and urea mix to a Microcon-10kDa Centrifugal Filter Unit and centrifuge at 14,000 ×g for 40 min at 20°C. Discard the flowthrough.

7. Add 200 µL of Wash buffer 1 and centrifuge at 14,000 ×g for 40 min at 20°C. Discard the flowthrough.

8. Add 100 µL of the alkylation solution and centrifuge at 14,000 ×g for 40 min at 20°C. Discard the filtrate.

9. Add 100 µL of Wash buffer 2 and centrifuge at 14,000 ×g for 40 min at 20°C. Repeat the wash with Wash buffer 2.

10. Dilute the protein concentrate by using 100 µL of 50 mM ammonium bicarbonate. This step is crucial for successful protease digestion which ultimately

determines the quality of the identification and quantification of peptides in mass spectrometry analysis. (Note 7)

11. Add trypsin at 1:20 enzyme to protein ratio, e.g., 5 μg of enzyme for 100 μg of protein sample.
12. Add 1 μL of 1% ProteaseMax Surfactant into the protein and trypsin mix and digest protein samples at 37°C for 3 h.
13. Transfer the centrifugal filter unit into a collection tube and spin at 14,000 ×g for 40 min (20°C). Keep the flowthrough which includes peptides.
14. Rinse the centrifugal filter unit using 50 μL of rinse solution and centrifuge at 14,000 ×g for 40 min at 20°C to prevent sample loss. (Note 8)
15. To acidify the peptides, add 6 μL of sample acidification buffer. (The final TFA concentration should be equal to 0.1%.)
16. Add 100 μL of wetting solution into the C18 tips and discard the flowthrough. Repeat x2.
17. Add 100 μL of equilibration buffer and discard the flowthrough. Repeat x2.
18. Transfer the peptide sample into the C18 tip and aspirate and dispense to provide peptide binding.
19. Add 100 μL of rinse solution and discard the flowthrough. Repeat x2.
20. Elute the peptides into a clean microcentrifuge tube by the addition of 50 μL of elution buffer. Repeat x2.
21. Transfer the peptide sample to a SpeedVac dryer and centrifuge at 48°C until they are completely dry. (Note 9)
22. Resuspend samples in LC/MS-compatible buffer. Samples are ready for LC-MS analysis.

4.3 PreOmics iST protocol

The PreOmics iST kit includes a LYSE buffer which provides denaturation, reduction, and alkylation of proteins at one step, significantly reducing the hands-on time during sample preparation. The DIGEST provided with the kit includes a mixture of Trypsin and LysC enzymes which provides cleavage of proteins at the carboxyl side of arginine and lysine residues. The recommended starting material by the manufacturer is between 1 and 100 μg. The methodology is described in detail in this section and a schematic of the workflow for the use of the PreOmics iST kit for mass spectrometry sample preparation from mammalian cell lines is provided in Fig. 3.3. The PreOmics iST kit uses protein starting materials extracted from cells, tissue, and bodily fluids. For the sample collection from mammalian cells, briefly, cells are grown in cell culture flasks with their respective media and when they reach confluency (80%−90%) they are trypsinized and centrifuged at 300 ×g for 5 min for collection of the cell pellet. The pellet is washed 3 times with PBS and PBS is removed after the final wash. The PreOmics iST kit can be used directly on the cell pellets including 1−100 μg of protein or pellets can be snap-frozen using liquid nitrogen and stored at −80°C until all samples are ready. It is important to note that, especially for label-free mass spectrometry−based proteomic analysis, all cell

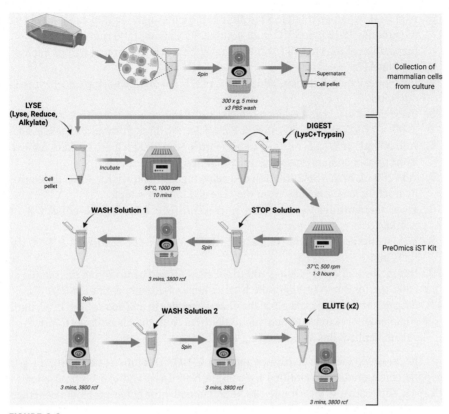

FIGURE 3.3

Workflow of collection of mammalian cells from cell culture and mass spectrometry sample preparation using PreOmics iST kit.

pellets should be collected prior to lysis and protein extraction. If the cell pellets contain more than 100 μg of protein starting material, cell pellets can be resuspended in PBS, their protein amount can be quantified and volumes including 100 μg of protein can be used for the PreOmics iST kit.

1. Add 50 μL of LYSE buffer to 1–100 μg of protein starting material obtained from cell pellets and heat at 95°C for 10 min with continuous mixing at 1000 rpm using a heating block. This step provides lysis of the cells, denaturation, reduction, and alkylation of the proteins within the sample. This step may produce pressure in the 1.5/2 mL microcentrifuge tubes, so care needs to be taken. (Note 10). Some samples may have excess DNA and this can be solved using a sonicator. (Note 11).

2. Cool down the sample to room temperature and transfer each sample to an iST cartridge placed on a waste tube (2 mL) via an adaptor.

3. Add freshly prepared DIGEST solution to each sample, lightly cap, and digest the proteins for 2 h at 37°C with continuous shaking (500 rpm). For the preparation of the DIGEST solution and alternative digestion durations, please see Note 12.

4. Following 2 h of digestion, add 100 μL of STOP solution to stop the enzymatic digestion.

5. Place the tubes/cartridges containing samples into a microcentrifuge and spin at 3800 ×g for 3 min. In this step, a complete flowthrough should be achieved.

6. Add 200 μL of WASH1 solution and centrifuge at 3800 ×g for 3 min to remove hydrophobic contaminants from the digested protein sample.

7. Add 200 μL of WASH2 solution and centrifuge at 3800 ×g for 3 min to remove hydrophilic contaminants from the digested protein sample.

8. Place the cartridge into a clean microcentrifuge tube for the collection of peptides.

9. To elute the peptides, add 100 μL of ELUTE solution and centrifuge at 3800 ×g for 3 min.

10. Repeat the elution by adding 100 μL of ELUTE and centrifuging the sample at 3800 ×g for 3 min bringing the total volume of eluted peptides to 200 μL.

11. Discard the cartridge and place the eluted peptides in a SpeedVac (at 45°C, until completely dry) to evaporate volatile buffers from the eluate prior to downstream applications or proceeding to LC-MS.

The dried peptides can be resuspended in LC-LOAD solution (aiming for a 1 g/L peptide concentration), quantified by using a NanoDrop and can be proceeded to LC-MS. Alternatively, the sample can be fractionated for better proteome coverage or can be labeled prior to mass spectrometry to increase the comparability of the samples and reduce machine time.

4.4 Upstream and downstream applications

4.4.1 Membrane fractionation

The protocol can be adapted to adherent cells, suspension cells, and frozen cell pellets. In case of the use of frozen cell pellets, carry out the wash step with cell wash solution prior to freezing.

1. Collect 5×10^6 cells (cell scraper/trypsinization for adherent cells, direct collection for suspension cells) and centrifuge the cell suspension at 300 ×g for 5 min.

2. Remove the supernatant containing growth medium and resuspend the cell pellet in 3 mL of cell wash solution.

3. Centrifuge at 300 ×g for 5 min, discard the supernatant, and resuspend the cell in 1.5 mL of the cell wash solution.

4. Centrifuge at 300 ×g for 5 min and remove the supernatant. (Note 13).

5. Add 7.5 μL of protease inhibitor into a 750 μL of permeabilization buffer (100X protease inhibitor (e.g., #87785, Thermo Scientific Halt Protease Inhibitor Cocktail, EDTA-Free (100X)).

6. Add 750 μL of permeabilization buffer to the cell pellet and mix it briefly by vortexing.

7. Incubate the cell suspension on a heating block for 10 min at 4°C with constant mixing at 1000 rpm for permeabilization.

8. Centrifuge the permeabilized cells at 16,000 ×g for 15 min and transfer the supernatant (containing cytoplasmic proteins) into a clean tube.

9. Add 5 μL of protease inhibitor into 500 μL of solubilization buffer.

10. Transfer the solubilization buffer containing protease inhibitor onto the cell pellet and mix by pipetting up and down.

11. Incubate the pellet for 30 min at 4°C with constant mixing (1000 rpm).

12. Centrifuge the solubilized sample at 16,000 ×g for 15 min at 4°C.

13. Transfer the supernatant containing solubilized membrane proteins into a clean tube.

14. Cytoplasmic and membrane fractions can be directly used for downstream applications or can be aliquoted and stored at −80°C until required.

15. For LC-MS/MS applications protein purification is highly recommended to remove the detergent, e.g., ReadyPrep 2-D Cleanup Kit (BIO-RAD).

4.4.2 High pH reversed-phase fractionation

1. To condition the spin columns provided by the Pierce High pH reversed-phase peptide fractionation kit, remove the protective white tip from the bottom of the spin column and place the column into a 2 mL microcentrifuge tube.

2. Spin the column at 5000 ×g for 2 min to remove 1:1 water: DMSO solution and to pack the resin.

3. Place the spin column into a clean 2 mL microcentrifuge tube and remove the cap.

4. Add 300 μL of ACN into the spin column, replace the cap and spin at 5000 ×g for 2 min. Discard the flowthrough.

5. Repeat the ACN wash (Step 4), centrifuge the spin column and discard the flowthrough.

6. Add 300 μL of 0.1% TFA solution into the spin column and centrifuge at 5000 ×g for 2 min.

7. Repeat the TFA wash step (Step 6). Discard the flowthrough. The conditioned column is ready for use.

8. Prepare the wash solution and elution solutions according to Table 3.1 described in Section 2.4.

9. Add 300 μL of 0.1% TFA solution to the protein digest containing 10−100 μg of peptides.

10. Place the conditioned spin column into a 2 mL clean microcentrifuge tube, remove the cap, and transfer 300 μL of the solution onto the column. Replace the cap.
11. Centrifuge the spin column at 3000 ×g for 2 min. Retain and label the filtrate as a "flowthrough" fraction.
12. Place the spin column into a fresh 2 mL microcentrifuge tube and add 300 μL of HPLC grade water, spin the column at 3000 ×g for 2 min, and collect the filtrate. Label as "wash" fraction. (Note 14)
13. Place the spin column into a new microcentrifuge tube and add 300 μL of the elution buffer 1.
14. Centrifuge at 3000 ×g for 2 min to collect fraction 1.
15. Continue collection of fractions by the addition of increasing concentrations of ACN (from fraction no. 1 to 10, by using elution buffers as described in Table 3.1).
16. Place the labeled microcentrifuge tubes containing flowthrough, wash and elution fractions into SpeedVac, and evaporate the solvent.
17. Resuspend the dried peptides in LC-MS suitable solution prior to MS analysis e.g., 0.1% formic acid.

4.4.3 Peptide purification and fractionation: strong cation exchange

1. Condition Strong Cation Exchange spin column by adding 400 μL of diluent solution and centrifuge at 2000 ×g for 5 min. Discard the flowthrough.
2. Add 200 μL of the diluent solution to the dried peptides for resuspension and transfer the peptide sample to the conditioned Strong Cation Exchange spin column. Centrifuge at 2000 ×g for 5 min.
3. Place the spin column into a clean microcentrifuge tube and add 200 μL of elution buffer 1 (concentration of KCl = 10 mM) and centrifuge at 2000 ×g for 5 min and collect the flowthrough.
4. Repeat this step by replacing the elution buffer with increasing concentrations of KCl and collect each flowthrough.

4.4.4 Chemical labeling of digested protein samples using TMT kit

1. Equilibrate the TMT label reagents to room temperature prior to use. (Note 15)
2. Add 41 μL of anhydrous ACN to a 0.8 mg vial of labeling reagent (Note 16) and dissolve the reagent for 5 min with occasional mixing and centrifuge the vial in a tabletop centrifuge to collect all the contents at the bottom of the vial.
3. Transfer the 100 μL protein digest (containing 25–100 μg peptide digest) into the label reagent vial containing 41 μL of the reagent (Note 17) and incubate the peptide digest with the labeling reagent for 1 h at room temperature.
4. To quench the labeling reaction, add 8 μL of 5% hydroxylamine into the vial containing the labeling reagent and peptide digest and incubate for 15 min.
5. Combine equal amounts of each labeled sample into a clean microcentrifuge tube and dry the labeled peptides to dryness using a SpeedVac.

6. Clean-up samples using peptide desalting spin columns, alternatively fractionate, prior to LC-MS. Peptide desalting spin columns enable sensitive analysis of the digested protein samples by removing of the hydrophilic contaminants, therefore reducing the signal suppression in mass spectrometry. Briefly, digested protein samples from mammalian cells are loaded onto a desalting spin column equilibrated with ACN and 0.1% TFA. The peptides bind to hydrophobic resin in the columns under acidic conditions and desalted by washing with acidified water at low-speed centrifugation. Desalted peptides are eluted by washing the column with 0.1% TFA and 50% ACN and centrifugation (3000 ×g for 1 min). For peptide fractionation, please see Section 4.4.2.

5. Strengths and limitations of FASP and PreOmics iST protocol

This section focuses on the strength and limitations of the techniques which researchers can encounter using these methods (Table 3.3). The FASP method shows high applicability in many proteomics research laboratories due to the ease of finding materials/reagents required for the preparation of protein digests. The FASP technique also shows high flexibility in the choice of lysis buffer and/or enzyme. The ultrafiltration device provides effective removal of SDS. This carries significant importance both in terms of the efficiency of digestion enzyme (SDS will have a negative impact) and mass spectrometry (detergent contamination). However, FASP has also limitations such as being time-consuming, requiring lengthy hands-on time and expertise of the researcher for sample preparation. The

Table 3.3 Strengths and limitations of FASP and PreOmics iST protocol.

Method	Strengths	Limitations
FASP	• Common materials and chemicals for proteomics laboratories • SDS-based lysis buffers can be used • Lysis buffer and enzymes can be chosen by the user	• Lengthy procedure • Lengthy hands-on time • Low reproducibility especially at lower protein amounts • Up to 100 μg protein loading
PreOmics iST	• Reproducible • Rapid procedure • Minimal hands-on time • Straightforward and easy sample preparation • Many samples can be prepared using a 96× kit	• Up to 100 μg max. protein loading • Requires the use of materials/reagents from a commercially available kit (although adaptable to other lysis buffers, optimization is required)

number of centrifugation steps and lengthy preparation can also lead to sample loss and low reproducibility within replicates.

The PreOmics iST kit provides a straightforward and rapid sample preparation with high reproducibility for mass spectrometry. However, the choice of "lyse" buffer and "lyse" enzyme is not as flexible as the other in-solution digestion methods since the protocol and its efficiency are tested via the use of the reagents provided by the commercially available kits. For alternative uses, further optimizations are needed.

Both techniques allow the digestion of a 100 μg protein sample as a starting material which limits their use for analysis of posttranslational modifications which need a high amount of protein samples (phosphorylation ~5 mg, ubiquitination ~20 mg, etc.).

It is also important to note that both techniques show compatibility with many upstream and downstream techniques such as membrane fractionation and TMT labeling.

6. Optimization and troubleshooting/notes

Note 1. Take caution while using SDS. SDS needs to be weighed using either a chemical fume hood or a dust mask. Amounts prepared should be adjusted according to the number of samples.

Note 2. Ammonium bicarbonate solution needs to be prepared a maximum of 24 h before digestion.

Note 3. Trypsin can be aliquoted and stored at −20°C for further use.

Note 4. If the protein concentration within the starting material is unknown, it is recommended to carry out a protein assay, such as BCA, to adjust the starting material in the range from 1 to 100 μg. PreOmics GmBH states that 6×10^5 HeLa cells provides approximately 100 μg of protein.

Note 5. The PreOmics GmBH kit offers 8× and 96× reaction size kits. In the materials and methodology, the protocol is described according to the 8× reaction size which uses collection and waste tubes (2 mL size tubes), whereas 96× reaction size uses 96-well plates as collection and waste plates.

Note 6. During tip sonication, place the sample tube on ice to prevent overheating.

Note 7. Although high concentrations of urea have been commonly used for protein denaturation in proteomics, it has been shown to lead to carbamylation of N-termini of proteins as well as amino groups on the side chain of lysine and arginine residues. Therefore, protein samples containing high concentrations of urea need to be diluted prior to digestion and mass spectrometry analysis.

Note 8. Eluted peptides can be stored at −80°C for further use.

Note 9. At this stage, dried peptide samples can be stored at −80°C or the procedure can be continued by proceeding to Strong Cation Exchange.

Note 10. This step may produce pressure in 1.5/2 mL microcentrifuge tubes, so care needs to be taken. This may result in the accumulation of sample droplets on the tube cap and sidewalls. The sample can be collected at the bottom of the tube by carrying out a short spin using a benchtop microcentrifuge.

Note 11. If the sample contains DNA, the lysed sample can be highly viscous. DNA can be sheared using a sonicator (10 cycles; 30 s ON/OFF).

Note 12. Add 210 µL of RESUSPEND solution into the DIGEST tube and mix at room temperature (500 rpm) for 10 min. One resuspended DIGEST tube provides enough digestion reagent for four samples. The manufacturer suggests digestion times between 1 and 3 h.

Note 13. Cells can be frozen and kept at −80°C.

Note 14. TMT labeled peptide samples require an additional step of washing with 5% ACN and 0.1% triethylamine for complete removal of unreacted labeling reagent. Please refer to Table 3.1, WASH for the preparation of the wash solution.

Note 15. TMT Isobaric label reagents are moisture sensitive. They need to be equilibrated to room temperature, immediately before use.

Note 16. For 0.2, 0.8, and 5 mg vials of TMT labeling reagents, adjust the amount of anhydrous ACN, e.g., 256 µL of anhydrous ACN for a 5 mg vial.

Note 17. Cell lysis, protein extraction, alkylation, reduction, and digestion can vary based on common lab practices. Thermo Fisher Scientific also provide a TMT labeling kit including the required reagents and solutions for these steps, e.g., TMT10plex Isobaric Mass Tag Labeling Kit, Catalog No: 90113.

7. **Alternative procedures and protocols**

There are many commercially available lysis buffers that have been developed and they are commercially available either on their own or as a part of the kit. It is always possible for the proteomic researcher to prepare their lysis buffer based on the application. One can easily use nonionic detergents such as Triton X-100, NP-40, Tween 20, Tween80, and Octyl glucoside or ionic detergents such as deoxycholate and SDS. It is also possible to use zwitterionic detergents such as CHAPS and CHAPSO. Based on the type of the detergent, improved extraction of certain proteins (such as cytoplasmic and membrane) can be achieved. However, it is important to note that, for sample preparation of MS-based proteomics, removal of these detergents is critical and must be achieved effectively. Based on the detergent type, the removal strategy must be changed and adapted to prevent the inactivation of the enzyme as well as contamination of the MS instrumentation. Acetone precipitation is usually recommended for the removal of NP-40, Tween 20, and Tween80, whereas CHAPS, CHAPSO, and SDS require the use of filtration/FASP techniques.

8. Safety considerations and standards

Proteomic techniques which provide effective lysis of mammalian cells and preparation of protein digests use organic solvents and detergents. Therefore, great care should be taken for the storage and handling of these chemicals. Personal protective equipment must be worn all the time and solvents should be handled in fume hoods. A person working with chemicals should be familiar with the Material Safety Data Sheet (MSDS) for all the chemicals which will be employed for the experiment. Waste disposal should also be considered prior to experiments.

Biological samples also require great caution to work with. Although the level of caution can vary from cell line to cell line, including possible genetic modifications on the cell lines and their origin, there are some common practices which need to be applied while working with them. Biological materials should always be handled inside a biological safety cabinet or in a fume hood and full personal protective equipment must be worn. All centrifugation steps of biological specimens should be carried out using only safety capped enclosures. All specimens which will be vortex-agitated also need to be contained in safety capped enclosures. All equipment and surfaces which come into contact with a biological material should be cleaned, disinfected with appropriate agents (depending on the biological sample and risk level) and/or autoclaved at the end of work. Safety documentation should be read carefully before handling any biological material.

9. Summary

In this chapter, we have provided more recent methodologies for mass spectrometry sample preparation from mammalian cell lines. Although the MS-based sample preparation techniques still have limitations, both FASP and the PreOmics iST provide an effective strategy by employing a "proteomic reactor" system. Both techniques can be adapted to upstream and downstream applications such as fractionation of cellular compartments and chemical labeling. The commercially available PreOmics iST kit provides additional advantages such as straightforward protocol and rapid sample preparation.

Acknowledgments

This work was supported and funded by a Science Foundation Ireland (SFI) Frontiers for the Future Award (grant no. 19/FPP/6759) and an Irish Research Council Postdoctoral Fellowship (Government of Ireland) award (grant no. GOIPD/2021/463) to Dr. Esen Efeoglu. Figs. 3.1, 3.2, and 3.3 were created using BioRender.

References

[1] Alberts B, Johnson A, Lewis J, Raff M, Roberts K, Walter P. Protein function. In: Molecular Biology of the Cell. 4th Edition; 2002. Available from: https://www.ncbi.nlm.nih.gov/books/NBK26911/.

[2] Deschênes-Simard X, Lessard F, Gaumont-Leclerc MF, Bardeesy N, Ferbeyre G. Cellular senescence and protein degradation. Cell Cycle June 15, 2014;13(12):1840—58.

[3] Nesvera J, Hochmannová J. DNA-protein interactions during replication of genetic elements of bacteria. Folia Microbiol 1985;30(2):154—76.

[4] Gonzalez MW, Kann MG. Chapter 4: protein interactions and disease. PLoS Comput Biol December 27, 2012;8(12):e1002819.

[5] Reynaud E. Protein misfolding and degenerative diseases. Learn Science at Scitable 2010. Available from: https://www.nature.com/scitable/topicpage/protein-misfolding-and-degenerative-diseases-14434929/.

[6] Proteomics — an overview. Science Direct Topics. Available from: https://www.sciencedirect.com/topics/neuroscience/proteomics.

[7] Meleady P. Two-dimensional gel electrophoresis and 2D-DIGE. In: Ohlendieck K, editor. Difference gel electrophoresis: methods and protocols. New York, NY: Springer; 2018. p. 3—14. https://doi.org/10.1007/978-1-4939-7268-5_1 (Methods in Molecular Biology).

[8] Viswanathan S, Ünlü M, Minden JS. Two-dimensional difference gel electrophoresis. Nat Protoc August 2006;1(3):1351—8.

[9] Macklin A, Khan S, Kislinger T. Recent advances in mass spectrometry-based clinical proteomics: applications to cancer research. Clin Proteom May 24, 2020;17(1):17.

[10] Yates JR. Recent technical advances in proteomics. F1000Research March 29, 2019;8:F1000. Faculty Rev-351.

[11] Feist P, Hummon AB. Proteomic challenges: sample preparation techniques for microgram-quantity protein analysis from biological samples. Int J Mol Sci February 5, 2015;16(2):3537—63.

[12] Chen EI, Cociorva D, Norris JL, Yates JR. Optimization of mass spectrometry compatible surfactants for shotgun proteomics. J Proteome Res July 2007;6(7):2529—38.

[13] Vere G, Kealy R, Kessler BM, Pinto-Fernandez A. Ubiquitomics: an overview and future. Biomolecules October 17, 2020;10(10):E1453.

[14] Manza LL, Stamer SL, Ham AJL, Codreanu SG, Liebler DC. Sample preparation and digestion for proteomic analyses using spin filters. Proteomics May 2005;5(7):1742—5.

[15] Wiśniewski JR, Zougman A, Nagaraj N, Mann M. Universal sample preparation method for proteome analysis. Nat Methods May 2009;6(5):359—62.

[16] Deeb SJ, Cox J, Schmidt-Supprian M, Mann M. N-Linked glycosylation enrichment for in-depth cell surface proteomics of diffuse large B-cell lymphoma subtypes. Mol Cell Proteomics MCP January 2014;13(1):240—51.

[17] Erde J, Loo RRO, Loo JA. Enhanced FASP (eFASP) to increase proteome coverage and sample recovery for quantitative proteomic experiments. J Proteome Res April 4, 2014;13(4):1885—95.

[18] Wiśniewski JR, Nagaraj N, Zougman A, Gnad F, Mann M. Brain phosphoproteome obtained by a FASP-based method reveals plasma membrane protein topology. J Proteome Res June 4, 2010;9(6):3280—9.

[19] Coleman O, Henry M, Clynes M, Meleady P. Filter-aided sample preparation (FASP) for improved proteome analysis of recombinant Chinese hamster ovary cells. Methods Mol Biol Clifton NJ 2017;1603:187—94.

[20] Kulak NA, Pichler G, Paron I, Nagaraj N, Mann M. Minimal, encapsulated proteomic-sample processing applied to copy-number estimation in eukaryotic cells. Nat Methods March 2014;11(3):319—24.

[21] Sielaff M, Kuharev J, Bohn T, Hahlbrock J, Bopp T, et al. Evaluation of FASP, SP3, and iST protocols for proteomic sample preparation in the low microgram range. J Proteome Res November 3, 2017;16(11):4060—72.

Sample preparation for proteomics and mass spectrometry from clinical tissue

Stephen Gargan[1,2], Paul Dowling[1,2] and Kay Ohlendieck[1,2]

[1]*Department of Biology, Maynooth University, National University of Ireland, Maynooth, Ireland;*
[2]*Kathleen Lonsdale Institute for Human Health Research, Maynooth University, National University of Ireland, Maynooth, Ireland*

1. Before you begin

The mass spectrometric analysis of human samples should be based on an optimized proteomic workflow for the unequivocal identification of proteins that are involved in pathogenic processes [1] and should ideally also take into account important ethical principles and a critical assessment of bioanalytical limitations [2]. Tissue proteomics can be carried out with freshly biopsied specimens or can be based on stored material in the form of freshly quick-frozen samples, optimal cutting temperature-embedded cellular material, or formalin-fixed and paraffin-embedded tissue [3]. This protocol describes the proteomic analysis of freshly dissected muscle biopsy material [4]. Most muscle biopsies are taken from *quadriceps*, *deltoid*, or *biceps* muscles [5], because extensive histological and histochemical knowledge has been assembled on these types of human skeletal muscles, such as size and distribution of different contractile fiber types [6]. A video link is available that describes in detail the percutaneous needle biopsy procedure to obtain a tissue sample from the *vastus lateralis* muscle [7]. The proteomic analysis of muscle biopsy specimens is crucial for studying the complex pathogenesis of neuromuscular disorders [8] such as sarcopenia of old age [9—11] or inherited muscular dystrophy [12,13]. Proteomic analysis of quick-frozen human muscle samples has been shown to result in an excellent coverage of the skeletal muscle proteome [14]. Prior to the preparation of tissue samples and the extraction of proteins, solutions and buffers should be freshly prepared. In order to avoid the potential degradation of chemicals in biological buffers, solutions should ideally be used the same day. It is crucial to have all essential pieces of equipment needed for the proteomic analysis of biopsy specimens, as listed in below key resources table, in good working order.

Proteomics Mass Spectrometry Methods. https://doi.org/10.1016/B978-0-323-90395-0.00011-5

1.1 Timing: 3–4 h

Buffers and solutions are required for biopsy preparation, tissue homogenization, the determination of protein concentration, the controlled digestion of extracted proteins and the removal of potentially interfering chemicals, as well as the liquid chromatographic separation of digested peptide populations and their subsequent mass spectrometric analysis. It is crucial to prepare all solutions with ultrapure and LC-MS compatible water and analytical grade/supra-pure reagents.

1. Biopsy buffer (phosphate-buffered saline, pH 7.4)
   (a) 0.137 M NaCl
   (b) 2.7 mM KCl
   (c) 10 mM Na_2HPO_4
   (d) 1.8 mM KH_2PO_4
   > Dissolve 8 g NaCl, 0.2 g KCl, 1.44 g Na_2HPO_4 and 0.25 g of KH_2PO_4 by stirring in dH_2O and bring to 1 L with dH_2O, and adjust pH to 7.4.
2. Tris buffer, pH 7.8
   (a) 0.1 M Tris
   (b) HCl (36%)
   > Dissolve 12.11 g of Tris base by stirring in dH_2O and bring to 1 L with dH_2O, and adjust pH to 7.8 with HCl.
3. Ammonium bicarbonate buffer
   (a) 50 mM ammonium bicarbonate
   (b) 0.1 M Tris buffer, pH 7.8
   > Dissolve 1.95 g of ammonium bicarbonate in Tris buffer, pH 7.8 and bring to 0.5 L with Tris buffer.
4. Sample homogenization buffer
   (a) 4% (w/v) sodium dodecyl sulfate
   (b) 0.1 M dithiothreitol
   (c) 50 mM ammonium bicarbonate buffer, pH 7.8
   > Dissolve 4 g of sodium dodecyl sulfate and 1.54 g of dithiothreitol in 50 mM ammonium bicarbonate buffer and bring to 100 mL with 50 mM ammonium bicarbonate buffer, pH 7.8.
5. Urea buffer
   (a) 8 M urea
   (b) 0.1 M Tris buffer, pH 8.5
   > Dissolve 120.12 g of urea in 0.1 M Tris buffer and bring to 0.25 L with 0.1 M Tris buffer, pH 8.5.
6. Iodoacetamide solution
   (a) 50 mM iodoacetamide
   (b) 8 M urea buffer
   > Dissolve 46 mg of iodoacetamide in urea buffer and bring to 5 mL with urea buffer.

7. Protein digestion buffer
   (a) Trypsin protease, MS-grade (50:1 protein:trypsin ratio)
   (b) 50 mM ammonium bicarbonate buffer, pH 7.8
   Dissolve 20 ng of MS-grade trypsin (per 1000 ng of protein) in ammonium bicarbonate buffer and bring to 5 mL with ammonium bicarbonate buffer.

8. MS sample buffer
   (a) 2% (v/v) trifluoroacetic acid
   (b) 20% (v/v) acetonitrile
   (c) LC-MS grade water
   Mix 1.0 mL trifluoroacetic acid with 10.0 mL acetonitrile and bring to 50 mL with dH_2O.

9. MS activation buffer
   (a) 50% (v/v) acetonitrile
   (b) LC-MS grade water
   Mix 10 mL acetonitrile with dH_2O and bring to 20 mL with dH_2O.

10. MS equilibrium/wash solution
    (a) 0.5% (v/v) trifluoroacetic acid
    (b) 5% (v/v) acetonitrile
    (c) LC-MS grade water
    Mix 0.25 mL trifluoroacetic acid with 2.5 mL acetonitrile and bring to 50 mL with dH_2O.

11. MS elution buffer
    (a) 80% (v/v) acetonitrile
    (b) LC-MS grade water
    Mix 4 mL of acetonitrile with dH_2O and bring to 5 mL with dH_2O.

12. MS resuspension buffer
    (a) 2% (v/v) acetonitrile
    (b) 0.1% (v/v) trifluoroacetic acid
    (c) LC-MS grade water
    Mix 0.1 mL acetonitrile with 5 μL trifluoroacetic acid and bring to 5 mL with dH_2O.

13. MS trapping buffer
    (a) 2% (v/v) acetonitrile
    (b) 0.1% (v/v) trifluoroacetic acid
    (c) LC-MS grade water
    Mix 20 mL of acetonitrile with 1 mL trifluoroacetic acid and bring to 1 L with dH_2O.

14. Solvent A for liquid chromatography
    (a) 0.1% (v/v) formic acid
    (b) LC-MS grade water
    Mix 1 mL formic acid with dH_2O and bring to 1 L with dH_2O.

15. Solvent B for liquid chromatography
    (a) 80% (v/v) acetonitrile
    (b) 0.08% (v/v) formic acid
    (c) LC-MS water

Mix 800 mL acetonitrile with 80 μL formic acid and bring to 1 L with dH_2O.

2. Key resources table

Reagent or resource	Source	Identifier
Biological samples		
Pathological human muscle biopsy/ autopsy specimens	Clinical samples	n/a
Unaffected human control biopsy specimens	Control samples	n/a
Chemicals, proteins		
Acetonitrile	Sigma	34851
Ammonium bicarbonate	Sigma	A6141
Bovine serum albumin	ThermoFisher Scientific	23208
Dithiothreitol	ThermoFisher Scientific	BP172-5
Formic acid	Sigma	5330020050
Hydrochloric acid	Merck	1.15186
Iodoacetamide	Acros Organics	122270050
Isopentane	Merck	PHR1661
LC-MS grade water	Sigma	39253
Liquid nitrogen	BOC Gases Ireland	n/a
Potassium chloride	Sigma	P9541
Potassium phosphate, monobasic	Sigma	P9791
Sodium chloride	Sigma	S3014
Sodium phosphate, dibasic	Sigma	S3264
Sodium dodecyl sulfate	Sigma	L3771
Trifluoroacetic acid	Sigma	T6508
Tris base	Sigma	T1503
Trypsin protease	ThermoFisher Scientific	90305
Urea	Sigma	U0631
Critical commercial assays		
Halt Protease Inhibitor Cocktail	ThermoFisher Scientific	78429
Ionic Detergent Compatibility Reagent for Pierce 660 nm Protein Assay Reagent	ThermoFisher Scientific	22663
Pierce 660 nm Protein Assay Reagent	ThermoFisher Scientific	1861426
Software and algorithms		
Progenesis QI for Proteomics	Waters Chromatography Ireland Ltd.	n/a
Proteome Discoverer 2.2 using Sequest HT	ThermoFisher Scientific	OPTON-30945

Reagent or resource	Source	Identifier
Other		
Analytical weighing scale	Farnell	3290043
Benchtop centrifuge	Eppendorf	5427R
Filter unit Vivacon 500	Sartorius	VN0H22
Heated electrospray ionization (H-ESI) ion source	ThermoFisher Scientific	H-ESI probe
Incubator	Memmert	INB200
Liquid nitrogen-cooled mini mortar plus pestle	ThermoFisher Scientific	H37260-0100
Manual single channel pipettes (20, 100, 1000 µL)	Mettler Toledo	17014391
Microplate reader	ThermoFisher Scientific	VL0000D0
Open Dewar flask for liquid nitrogen	BOC Gases Ireland	11880462
Orbitrap Fusion Tribrid mass spectrometer	ThermoFisher Scientific	IQLAAEGAAPFADBMBCX
Pierce C18 spin columns	ThermoFisher Scientific	89870
Polypropylene micro pellet pestles for plastic tubes	Thomas Scientific	3411E25
Reverse-phased capillary high-pressure liquid chromatography system	ThermoFisher Scientific	UltiMate 3000 HPLC
Safe-lock plastic tubes for sample storage	Eppendorf	0030121597
Sonicator	Bandelin	UW2200
Thermomixer	Eppendorf	5382000031
Vacuum evaporator	Genevac	DNA-12060-C00
Vortex	Sigma	Z258423

3. Materials and equipment

- Sterile medical gauze for transportation of fresh muscle biopsy material
- Analytical weighing scale
- Phosphate-buffered saline for preparation of biopsy samples
- Liquid nitrogen storage facility
- Open Dewar flask and metal beaker for quick-freezing of biopsy material in liquid nitrogen-cooled isopentane
- Liquid nitrogen—cooled mini mortar plus pestle
- Sample homogenization buffer for resuspension of pulverized biopsy material
- Commercial protein assay system to determine protein concentration using bovine serum albumin or another suitable protein as standard
- Buffers for protein preparations (Tris buffer, ammonium bicarbonate buffer, urea buffer, iodoacetamide solution)

- Suitable filter units for filter-aided sample preparation, such as Vivacon 500 (10000 MWCO)
- Thermomixer
- Trypsin-containing protein digestion buffer
- Benchtop centrifuge
- Incubator
- Manual single channel pipettes
- Microplate reader
- Pierce C18 spin columns
- Polypropylene micro pellet pestles for plastic tubes
- Safe-lock plastic tubes for sample storage
- Sonicator
- Vacuum evaporator
- Vortex
- MS sample buffer
- MS activation buffer
- MS equilibrium/wash solution
- MS elution buffer
- MS resuspension buffer
- MS trapping buffer
- Solvent A for liquid chromatography
- Solvent B for liquid chromatography
- Reverse-phased capillary high-pressure liquid chromatography system
- Heated electrospray ionization (H-ESI) ion source
- Mass spectrometer (Orbitrap Fusion Tribrid MS apparatus)
- Software (Progenesis QI for Proteomics; Proteome Discoverer 2.2 using Sequest HT)

Alternatives: This protocol describes the proteomic identification of proteins extracted from skeletal muscle biopsy specimens using liquid chromatographic separation combined with analysis by an Orbitrap Fusion Tribrid mass spectrometer from Thermo Fisher Scientific. Alternatively, a variety of other proteomic methods, mass spectrometers and proteomic analysis software packages can be employed for the systematic identification and characterization of proteins extracted from clinical tissue specimens [1]. The determination of protein concentration described here is carried out with the help of a Pierce 660 nm Protein Assay system. Alternatively, various other commercially available protein assays can be used to measure the amount of protein being present in fractions extracted from clinical tissue samples.

4. Step-by-step method details

4.1 Preparation of biopsy specimens

4.1.1 Timing: 1–2 h

1. Keep skeletal muscle tissue samples, which were received from diagnostic biopsies, operational remnants, or autopsy material, in their original state for transportation. Specimens should initially not be immersed in any buffer or solution [5].

2. Properly label all sampling containers for the transportation or storage of individual tissue samples. The labeling system should be adopted to accommodate the analysis of individual samples or multiplexing approaches to study large sets of tissue specimens.

3. Strictly follow national regulations and ethical guidelines if anonymization of human tissue specimens is required [2].

4. Transport freshly dissected muscle samples in sterile medical gauze following extraction by open biopsy or a needle biopsy procedure [15].

5. Carefully remove excess fat and connective tissue, if necessary, by physical means from the muscle specimen.

6. Determine the wet weight of the biopsy material with an analytical weighing scale.

7. Depending on whether the biopsied tissue specimens will be used for both histological/histochemical examinations and biochemical analyses or only for proteomic studies, the samples should be prepared differently, as outlined below.

8. The initial handling of fresh biopsy specimens and their preparation for proteomic analysis versus long-term storage for subsequent studies is outlined in Fig. 4.1:

    (a) If tissue specimens are exclusively used for biochemical studies, the sample can be briefly washed with ice-cold biopsy buffer, such as phosphate-buffered saline at pH 7.4, to remove excess blood.

    (b) Importantly, if the biopsy material was dissected for both histological analyses and proteomic surveys, samples should not be immersed in any solution or buffer. In this case, tissue specimens should be immediately quick-frozen in liquid nitrogen—cooled isopentane.

9. For quick-freezing of muscle samples for proteomic analysis, carefully fill an open 2 L Dewar flask with liquid nitrogen and insert a small metal container with an isopentane solution.

10. The metal container should easily fit into the larger Dewar flask and be fixed in position by hanging from a metal rod over the opening of the open Dewar flask.

11. Fill the metal container approximately half-full with the isopentane solution and slowly suspend it into the liquid nitrogen—filled Dewar container.

12. During the initial cooling process of the isopentane solution, white aggregates will start to appear at the edges of the container, often referred to as "caking." This is the optimum period for introduction of the biopsy material into the cooled isopentane solution.

13. Once the tissue sample is thoroughly frozen, it can be removed and transferred to a labeled and precooled plastic safe-lock tube.

Note: Avoid the immersion of freshly dissected muscle biopsy material in solutions if it is planned to use the sample for both the microscopical assessment of transverse cryosections and mass spectrometry-based proteomic analysis. The presence of liquids may cause freezing artifacts in skeletal muscle specimens and should therefore be avoided during the initial preparation, transportation, and storage for the subsequent diagnostic usage in the form of histological and histochemical staining procedures [4].

Critical: Excess blood, fat, and connective tissue are routinely removed from skeletal muscle biopsies prior to diagnostic procedures. However, if muscle fiber populations are supposed to be studied by proteomics in their entire tissue environment to evaluate necrosis, fibrosis, inflammation, fat substitution, or other pathological changes, it is critical to skip this preparative step and take the entire biopsy material as starting material for biochemical and mass spectrometric analysis.

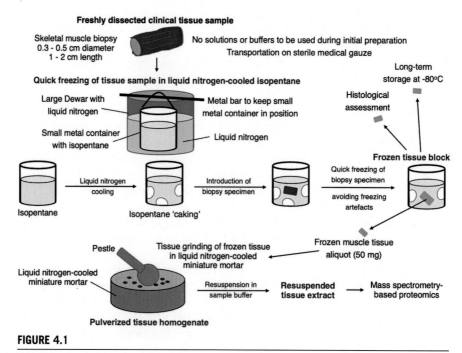

FIGURE 4.1

Initial handling and preparation of fresh muscle biopsy specimens for proteomic analysis.

Pause Point: If frozen biopsy samples do not have to be immediately processed for diagnostic procedures or mass spectrometric analysis, they can be transferred to a $-80°C$ freezer or a large liquid nitrogen containing Dewar container for long-term storage. Properly quick-frozen skeletal muscle specimens obtained by biopsy procedures or sampling during autopsy can be stored for many years without significant degradation of proteins [14].

4.2 Homogenization of tissue material
4.2.1 Timing: 2–3 h

1. The overall analytical workflow of the mass spectrometry—based proteomic analysis of clinical tissue samples is summarized in Fig. 4.2. For the proteomic analysis of small amounts of tissue material from needle biopsies (50 mg), pulverization of frozen tissue samples is often advantageous over the usage of mechanical tissue homogenizers [16] and usually gains a satisfactory yield of extracted proteins:

   (a) Prior to homogenization, the weight of the pulverized tissue sample should be determined with an analytical weighing scale.

   (b) Transfer the frozen biopsy material into a suitably sized mortar. If a conventional mortar is used, the device should be precooled with liquid nitrogen.

   (c) For optimum grinding of tissue samples, ideally a commercially available liquid nitrogen—cooled miniature mortar should be employed.

   (d) Carefully grind the tissue material with the help of a cooled pestle until a fine powder is created. It is crucial to keep the system properly cooled with liquid nitrogen so that the tissue material does not thaw during the homogenization process.

   (e) Transfer the frozen tissue powder to a small and precooled plastic tube. For small samples, 1.5 mL microcentrifuge tubes are suitable for the next homogenization step.

   (f) Add ice-cold sample homogenization buffer to the tissue powder in a 1.5 mL microcentrifuge tube.

   (g) For proteomic studies, the ratio of extracted skeletal muscle tissue to homogenization buffer should be 1:6 (mg/µL).

   (h) Carefully mix this solution and resuspend the protein pellet with a polypropylene micropellet pestle.

2. Heat the suspended tissue extract for 5 min at 95°C.

3. If the suspension is excessively viscous, which can be the case with multinucleated contractile fibers following homogenization, large DNA molecules can be treated by sonication. Usually, four bursts of sonication lasting approximately 5 s sufficiently reduce the viscosity of muscle homogenates.

4. Centrifuge the suspended muscle biopsy material for 10 min at $20,000 \times g$ in a benchtop centrifuge.

5. Use the supernatant fraction for biochemical and proteomic analyses.

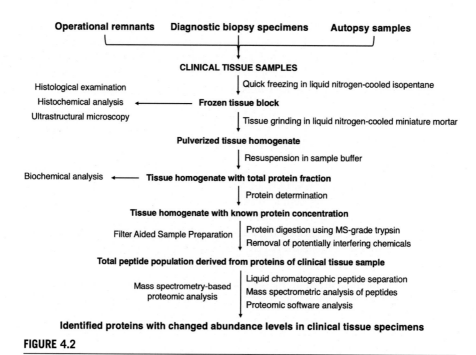

FIGURE 4.2

Workflow of the mass spectrometry—based proteomic analysis of clinical tissue samples.

Note: If additional biochemical studies are planned, such as gel electrophoresis or immunoblot analysis, the homogenization buffer should be supplemented with a protease inhibitor cocktail to avoid the proteolytic degradation of sensitive skeletal muscle proteins.

Critical: During all preparative steps involved in the harvesting and preparation of potentially infectious clinical tissue samples, as well as handling of liquid nitrogen and potentially harmful chemicals such as sodium dodecyl sulfate and iodoacetamide, a double layer of protective gloves, safety goggles, face mask and a laboratory coat should be worn.

Pause Point: Following tissue grinding and resuspension, the cleared supernatant fraction containing biopsy extracts can be quick-frozen in liquid nitrogen and stored in a −80°C freezer until further processing.

4.3 Determination of protein concentration

4.3.1 Timing: 1—2 h

1. Prior to mass spectrometric analysis, the protein concentration of the supernatant fraction containing extracted biopsy material should be determined [17].
2. Use bovine serum albumin as a protein standard. A standard graph can be established with 25, 50, 125, 250, 500, 750, 1000, 1500, and 2000 µg protein per mL.

3. Transfer 10 μL of each bovine serum albumin protein standard to individual microplate wells in triplicate.
4. Transfer 10 μL of the extracted biopsy samples with unknown protein concentration to individual wells on the same microplate in triplicate.
5. To zero the plate reader, a blank sample is also needed. Transfer 10 μL of a blank sample, such as the homogenization buffer into individual wells on the same microplate in triplicate.
6. If a commercial assay system is used, such as the Pierce 660 nm Protein Assay system [18], add 150 μL of the protein assay reagent that is supplemented with ionic detergent compatibility reagent to each well in the microplate.
7. The plate should be covered and then carefully mixed on a plate shaker at medium speed. Shaking at approximately 600 rpm for 1 min is usually sufficient for the mixing of the protein samples and the dye reagent.
8. Incubate the mixture of the protein suspension and the assay reagent for 5 min at 20°C.
9. Zero the plate reader by using the blank wells as baseline values.
10. To establish a standard graph with bovine serum albumin, measure the absorbance of the various standards ranging from 50 to 2000 μg protein per mL at a wavelength of $\lambda = 660$ nm.
11. Plot the average blank-corrected 660 nm measurement for each protein standard versus its concentration in μg/mL to prepare a standard curve.
12. Measure the absorbance of the biopsy samples with unknown protein concentration at a wavelength of $\lambda = 660$ nm.
13. Compare the absorbance values of the biopsy samples with the standard curve to determine the protein concentration of each unknown sample.

Note: To avoid potential issues with linearity across a limited range of concentration magnitudes, a four-parameter curve fit should be used.

Critical: The sample homogenization buffer contains 1% (w/v) sodium dodecyl sulfate, which can interfere with protein concentration measurements. It is therefore critical to add ionic detergent compatibility reagent to the protein assay reagent.

4.4 Digestion of protein sample

4.4.1 Timing: 10–24 h

1. Transfer the sample homogenate containing 25 μg protein and 200 μL of urea buffer to a centrifugal filter unit, such as the Sartorius Vivacon 500 unit.
2. Briefly vortex the suspension.
3. Centrifuge the filter unit for 15 min at 14,000 × g.
4. Add another 200 μL of urea buffer to the filter unit.
5. Briefly vortex the suspension.
6. Centrifuge the filter unit for 15 min at 14,000 × g.
7. Discard the flowthrough solution.
8. Add 100 μL of iodoacetamide solution and cover the filter units in tinfoil.

9. Mix the solution for 1 min at 600 rpm in a thermomixer.
10. Incubate the solution without mixing for 20 min in the dark.
11. Centrifuge the filter unit for 15 min at 14,000 × g.
12. Add 100 μL of urea buffer to the filter unit.
13. Briefly vortex the suspension.
14. Centrifuge the filter unit for 15 min at 14,000 × g.
15. Add another 100 μL of urea buffer to the filter unit.
16. Briefly vortex the suspension.
17. Centrifuge the filter unit for 15 min at 14,000 × g.
18. Add 100 μL of ammonium bicarbonate buffer.
19. Briefly vortex the suspension.
20. Centrifuge the filter unit for 15 min at 14,000 × g.
21. Add another 100 μL of ammonium bicarbonate buffer.
22. Briefly vortex the suspension.
23. Centrifuge the filter unit for 15 min at 14,000 × g.
24. Transfer the centrifugal filter units to fresh collection tubes.
25. Add 40 μL of digestion buffer containing trypsin (50:1 protein:protease ratio) and keep this solution on ice.
26. Mix the solution for 1 min at 600 rpm in a thermomixer.
27. The filter units containing the protein sample and trypsin should then be positioned in a suitable wet chamber.
28. Incubation should be carried out for 4–18 h at 37°C in a sterile incubator.
29. Following the digestion step, the filter units should be placed in new collection tubes.
30. Centrifuge the filter unit for 10 min at 14,000 × g.
31. The filtrate contains the generated peptide fraction.
32. Add 40 μL of ammonium bicarbonate buffer to the filter unit.
33. Centrifuge the filter unit for another 10 min at 14,000 × g.
34. The filtrate contains the combined peptide fraction.
35. Transfer 60 μL of each sample to new plastic tubes.
36. Add 15 μL of MS sample buffer, which contains 2% (v/v) trifluoroacetic acid and 20% (v/v) acetonitrile.
37. The final peptide sample solution should contain 0.5% trifluoroacetic acid in 5% acetonitrile.

Note: If incubation with trypsin does not result in a sufficient rate of proteolysis, other types of proteases or combinations of proteolytic enzymes can be used for the consecutive treatment of proteins [19].

Critical: The digestion step should be carried out in a wet chamber environment, which can be conveniently constructed with a sterile pipette box containing dH$_2$O and soaked tissue sheets. Sample tubes undergoing protein digestion can be held in a foam floater [20].

Pause Point: Following digestion of protein samples, the generated peptide solutions can be quick-frozen in liquid nitrogen and stored in a −80°C freezer until further processing.

4.5 Removal of interfering chemicals from peptide fraction

4.5.1 Timing: 1−2 h

1. For the simultaneous processing of multiple clinical samples, porous C18 tubes are placed into their respective receiver tubes.
2. Rinse the walls of C18 spin tubes and wet the reverse-phase resin by the addition of 200 μL of activation solution.
3. Centrifuge the C18 tubes for 1 min at 1500 × g.
4. Discard the flowthrough solution.
5. Add another 200 μL of activation solution.
6. Recentrifuge the C18 tubes for 1 min at 1500 × g.
7. Discard the flowthrough solution.
8. Add 200 μL of equilibration/wash solution.
9. Centrifuge the C18 tubes for 1 min at 1500 × g.
10. Discard the flowthrough solution.
11. Add another 200 μL of equilibration/wash solution.
12. Recentrifuge the C18 tubes for 1 min at 1500 × g.
13. Discard the flowthrough solution.
14. Transfer the sample on top of the resin bed.
15. Place the C18 spin tubes containing the samples into receiver tubes.
16. Centrifuge the C18 tubes for 1 min at 1500 × g.
17. Recover the flowthrough fraction.
18. Pipette the flowthrough fraction on top of the resin bed.
19. Replace the C18 spin tubes into receiver tubes.
20. Recentrifuge the C18 tubes for 1 min at 1500 × g.
21. Transfer C18 spin tubes into new receiver tubes.
22. Add 200 μL of equilibration/wash solution to C18 spin tubes.
23. Centrifuge the C18 tubes for 1 min at 1500 × g.
24. Discard the flowthrough solution.
25. Add another 200 μL of equilibration/wash solution to C18 spin tubes.
26. Recentrifuge the C18 tubes for 1 min at 1500 × g.
27. Discard the flowthrough solution.
28. Transfer the C18 spin tubes into new receiver tubes.
29. Add 20 μL of elution buffer to the top of the resin bed.
30. Centrifuge the C18 tubes for 1 min at 1500 × g.
31. Add another 20 μL of elution buffer to top of the resin bed using the same receiver tubes.
32. Recentrifuge the C18 tubes for 1 min at 1500 × g.
33. Dry the eluted and combined fraction in a vacuum evaporator.
34. Resuspend the sample in 50 μL of resuspension buffer.

Note: Following resuspension, the final solution containing the digested and washed peptide fragments should contain 2% (v/v) acetonitrile and 0.1% (v/v) trifluoroacetic acid for transfer to the trapping column of a liquid chromatography system.

Critical: To avoid potential complications due to detergents with the subsequent mass spectrometric analysis, interfering chemicals are removed by washing and centrifugation of the resin-associated peptide fraction using commercially available spin columns prior to elution and separation by liquid chromatography [21]. Importantly, to ensure complete binding of sample protein, the flowthrough solution resulting from the initial incubation of washed C18 tubes is readministered to the resin and incubated for a second time, as described above.

Pause Point: Resuspended peptide solutions can be quick-frozen in liquid nitrogen and then stored in liquid nitrogen or a −80°C freezer for future analysis in a mass spectrometer.

4.6 Mass spectrometric LC-MS/MS analysis
4.6.1 Timing: 12–72 h

1. Use a suitable liquid chromatography system for optimum peptide separation, such as the Thermo Fisher Scientific UltiMate 3000 UHPLC system.
2. Load a 2 μL aliquot of peptides, generated by trypsin digestion and cleared of interfering chemicals, onto the trapping column, such as a PepMap100 (C18, 300 μm × 5 mm) column from Thermo Fisher Scientific.
3. Run the trapping column at a flow rate of 25 μL/min with trapping buffer for 3 min.
4. Following the application to the trapping column system, the sample is resolved by an analytical column, such as an Acclaim PepMap 100 (75 μm × 50 cm, 3 μm bead diameter) column.
5. Elute peptides with a binary gradient as follows: LC Solvent A and LC Solvent B using 2%–32% Solvent B for 75 min, 32%–90% Solvent B for 5 min and holding at 90% for 5 min at a flow rate of 300 nL/min.
6. Select a data-dependent acquisition method for the identification of peptides using a voltage of 2.0 kV and a capillary temperature of 320°C.
7. Perform data-dependent acquisition with full scans in the 380–1500 m/z range using a mass analyzer, such as the Thermo Fisher Scientific Orbitrap Fusion Tribrid.
8. Use a resolution of 120,000 (at m/z 200), a targeted automatic gain control (AGC) value of 4E+05 and a maximum injection time of 50 ms.
9. The top-speed acquisition algorithm determines the number of selected precursor ions for peptide fragmentation.
10. Isolation of selected precursor ions is performed in the quadrupole unit with an isolation width of 1.6 Da.
11. Peptides with a charge state ranging from 2+ to 7+ are analyzed and a dynamic exclusion is applied after 60 s.
12. Fragmentation of precursor ions is carried out with higher energy collision-induced dissociation using a normalized collision energy of 28%.
13. Generated MS/MS ions are measured in the linear ion trap using a targeted AGC value of 2E+04 and a maximum fill time of 35 ms.

Note: Prior to mass spectrometric analysis, an important step in the proteomic workflow is the separation of peptides by reversed-phase liquid chromatography [22]. A variety of high-performance liquid chromatography systems are commercially available. In this protocol, ultra high-performance liquid chromatography was carried out with the Thermo Scientific UltiMate 3000 UHPLC system. Alternatively, many other excellent chromatographical systems are available for proteomic applications. If liquid chromatography is carried out in a multiuser core facility, make sure that (1) columns are properly washed, (2) there is no sign of column deterioration, (3) the system is not contaminated by previous applications, (4) buffers are freshly made, and (5) the separation system is in good working order.

Critical: If peptide-centric approaches are employed for the unequivocal identification of individual proteoforms from clinical tissue specimens, it is crucial to use an optimized enzymatic or chemical protocol for controlled protein digestion. If the routine application of trypsin for proteolysis does not result in the satisfactory production of characteristic peptide populations, as evidenced by the pure mass spectrometric identification of proteins, then alternative digestion methods should be tested.

4.7 Protein identification

4.7.1 Timing: 2–3 h

1. Search the generated mass spectrometric files (.raw) against the *Homo sapiens* UniProtKB-SwissProt database with Proteome Discoverer 2.2 using Sequest HT and Percolator.
2. Locate the proteome data for "human" by searching by name or by taxonomy ID, i.e., *Homo sapiens* at: https://www.uniprot.org/proteomes/.
3. Click on the Proteome ID link - UP000005640 for "human" (78,120 proteins).
4. Select the most suitable database in UniProt, such as the reviewed (UniProtKB/ Swissprot), the unreviewed (UniProtKB/TREMBL) or the combined (UniProtKB) option.
5. Click on the Download button and choose: "All protein entries, FASTA (Canonical and isoform), compressed."
6. Use suitable search parameters for protein identification, including settings for peptide mass tolerance, MS/MS mass tolerance, the number of missed peptide cleavages, carbamido-methylation of cysteines and methionine oxidation:
   (a) Set the peptide mass tolerance to 10 ppm.
   (b) Set MS/MS mass tolerance to 0.6 Da.
   (c) Allow up to two missed peptide cleavages.
   (d) Set carbamido-methylation of cysteine as a fixed modification.
   (e) Set methionine oxidation as a variable modification.
7. Consider only highly confident peptide identifications.
8. Aim for a false discovery rate of FDR ≤ 0.01, identified using a SEQUEST HT workflow coupled with Percolator validation in Proteome Discoverer 2.2.

Note: A variety of commercially available software programs can be employed to analyze mass spectrometric files. Reviewed databases contain highly curated entries with a minimal level of redundancy and the ability to efficiently integrate with other databases for the identification of human proteoforms.

Critical: In order to carry out reliable protein identifications, it is important to define suitable search parameters in relation to peptide mass tolerance, MS/MS mass tolerance and the number of allowable missed peptide cleavages, as well as settings for the carbamido-methylation of cysteines and methionine oxidation.

4.8 Comparative profiling of protein abundance

4.8.1 Timing: 12–48 h

1. Import mass spectrometric raw files into a suitable software analysis program, such as Progenesis QI for Proteomics (Waters), for the comparative abundance profiling of proteins in normal controls versus the diseased state.
2. Run an automatic alignment to combine and compare the result from different LC-MS runs.
3. Carry out an automatic peak picking and matching across all data files.
4. Create an aggregate data set from the aligned runs, which contains all peak information from all sample files and allows the detection of a single map of peptides.
5. Apply this map to each individual sample, which then allows 100% matching of peaks with no missing values.
6. Normalize the peptide ion abundance measurements to allow the comparisons between the normal controls and the pathological specimens form clinical tissue specimens in order to identify peptides of pathobiological interest.
7. Base the determination of the peptide ions of interest on the significance measure of ANOVA with a P-value of 0.05.
8. Export MS/MS spectra from these peptides and carry out identifications by using the above-described process with Proteome Discoverer 2.2 software.
9. Reimport the result file into Progenesis QI for Proteomics, which then allows a detailed review of all peptide ions used to quantify and identify individual proteoforms.
10. Finally, base the determination of the human protein species of interest on the significance measure of ANOVA with a P-value of 0.05.

Note: A variety of commercially available software analysis programs can be utilized to carry out the comparative proteomic profiling of different protein fractions. In this protocol, Progenesis QI for Proteomics (Waters) software was used. User guides and tutorial data sets for Progenesis QI are available online: (http://www.nonlinear.com/progenesis/qi-for-proteomics/v4.0/user-guide/). The proteomic comparison of extracts from clinical tissue preparations isolated from normal control specimens versus patient specimens can lead to the identification of large numbers of differentially expressed proteins in human disease [12].

Critical: Prior to finalizing the list of significant human proteoforms that exhibit a significantly changed abundance in a particular disorder, it is important to review crucial quality control metrics in relation to sample preparation, instrumentation, and experimental parameters. The review of sample preparation metrics should address potential problems with the isolation and preparation of protein extracts for mass spectrometric analysis, such as issues with protein yield and/or a suboptimal digestion efficiency for the production of characteristic peptide populations. The critical assessment of instrument metrics should exam the suitability and configuration of the liquid chromatography system and mass spectrometer for optimum bioanalytical performance. In addition, a review of experiment metrics should ideally investigate the scale of protein identification and review the mass spectrometric identification of statistical outliers in an individual sample set.

5. Expected outcomes

The expected yield of total protein from clinical tissue samples to be used for proteomic studies depends on the amount and quality of starting material and the optimization of the protocol used for tissue harvesting, dissection, preparation, and storage of tissue specimens prior to mass spectrometric analysis. In contrast to relatively large amounts of tissue that can be obtained from open biopsy procedures, operational remnants or autopsy samples, which might result in several grams of wet weight, only small amounts of human muscle specimens can be routinely obtained by the less invasive and most frequently used needle biopsy method [23]. Diagnostic procedures and testing regimes during clinical trials use different types of biopsy procedures, which depend heavily on the anticipated range of histological, ultrastructural and/or biochemical tests. Most procedures produce cylindrical muscle samples of 0.3–0.5 cm in diameter and 1–2 cm in length of approximately 0.1–0.5 mg wet weight [5,15]. Crucial aims for the conduction of needle biopsy techniques include toleration of the procedure by both pediatric and adult patients, minimum interference with pharmacological substances such as anesthetics, technical reliability, patient safety, and the harvesting of sufficient amounts of high-quality tissue for routine histological and histochemical analyses. For additional pathobiochemical investigations, such as enzyme testing, immunoblotting, or mass spectrometry–based proteomic screening of clinical tissue samples, usually 50 mg of muscle tissue can be made available from patient biopsy material [5]. The expected outcome of a mass spectrometric analysis depends heavily on the proteomic approach, such as targeted versus discovery studies or bottom-up versus top-down proteomics, and the sensitivity of the mass spectrometer used for protein identification [24–26]. The usage of an Orbitrap type mass spectrometer usually results in the identification of several thousand proteoforms. Proteomic studies with crude skeletal muscle biopsy material cover routinely proteins involved in excitation-contraction coupling, ion homeostasis, the contraction–relaxation cycle, the cellular stress response, metabolite transportation, energy metabolism, cellular signaling cascades, cytoskeletal networks, and the various layers of the extracellular matrix [12].

6. Quantification and statistical analysis

The usage of sufficient technical and biological repeats during biochemical assays and proteomic surveys, as well as suitable software analysis programs and proper search parameters for protein identification, is essential for the successful conduction of a proteomic survey of clinical tissue specimens. Following protein extraction, the determination of the protein concentration of the unknown protein suspension and the absorbance values of the bovine serum albumin protein standards should be carried out in triplicate. In order to prevent issues with linearity across a limited range of protein concentration magnitudes, use a four-parameter curve fit. For the analysis of protein species derived from human tissue specimens, search the generated mass spectrometric files against a suitable databank, such as the *Homo sapiens* UniProtKB-SwissProt database with Proteome Discoverer 2.2 using Sequest HT and Percolator. It is crucial to only consider highly confident peptide identifications. The aim for a false discovery rate should be FDR ≤ 0.01 using a SEQUEST HT workflow coupled with Percolator validation in Proteome Discoverer 2.2. For the comparative profiling of protein abundance, the determination of the peptide ions of interest should be based on the significance measure of ANOVA with a *P*-value of 0.05. The final determination of the human protein species of interest should also be based on the significance measure of ANOVA with a *P*-value of 0.05. Ideally, the identification of a particular proteoform should be based on sequence information from at least two peptides.

7. Advantages

Mass spectrometry—based proteomics has developed into a robust and reliable bioanalytical tool for the large-scale identification of peptides and proteins of clinical interest [2]. In contrast to focusing on individual proteins and their potentially preconceived pathophysiological role, the great advantage of proteomic surveys is the unbiased and technology-driven approach that can result relatively swiftly in the comprehensive establishment of proteome-wide changes in a particular disease. If used as part of clinical studies, proteomics can be employed for the systematic discovery of novel biomarker candidates for improving differential diagnosis, prognosis, and therapy monitoring.

8. Limitations

Limitations of the proteomic analysis of extracts from clinical tissue specimens are related to potential bioanalytical issues with sample handling, mass spectrometry, single timepoint assaying, sample heterogeneity, and interindividual differences within patient populations. The systematic mass spectrometric survey of clinical tissue samples includes a variety of critical steps that may affect the quality of the proteomic data, including the method of surgical resection, as well as the initial preparation, transportation, and potential preservation of biopsy material. Especially critical can be complications due to extended delays between initial sample retrieval and the start of tissue homogenization and subsequent biochemical analysis. This may result in the degradation of sensitive protein species and can thereby potentially introduce bioanalytical artifacts. It is also important to remember that the entirety of the diverse protein constituents of a specific cellular arrangement represents a highly dynamic system with constant changes in protein abundance, isoform expression patterns, protein interactions, and posttranslational modifications. The cellular proteome in the human body is in a constant mode of adaptation to altered functional demands on the physiological, biochemical, metabolic, and signaling level [27]. Hence, the pathoproteomic profile of a particular clinical biopsy sample reflects only a single timepoint of protein expression at a specific disease stage. This issue should be taken into account when complex pathological protein changes are interpreted. In addition, the heterogeneity of human tissue specimens and considerable interindividual genetic differences between patients require usually larger numbers of samples for reliable statistical analysis as compared to studies with cell cultures or animal models of human disease.

9. **Optimization and troubleshooting**

9.1 **Issues with small amounts of clinical biopsy material**

The availability of sufficient amounts of starting material for the extraction of appropriate quantities of protein to be analyzed by mass spectrometry is a crucial factor for a successful proteome-wide identification of changes in distinct protein species. Small amounts of tissue are difficult to homogenize in a reproducible manner.

9.2 **Potential solution to optimize the procedure**

Ideally at least 25 mg wet weight of individual tissue samples should be used as starting material for proper homogenization by grinding with a liquid nitrogen—cooled miniature mortar. In the case of biopsy sampling in the field of neuromuscular disorders, often more than one muscle specimen is taken per patient, so aliquots from several tissue samples originated from the same individual can be pooled for the reproducible extraction of total tissue protein. However, if proteomic studies are not employed for diagnostic, prognostic, or therapy-monitoring purposes and the determination of altered protein expression levels or posttranslational modifications at the level of individual patients, but in order to conduct novel basic research into pathoproteomic mechanisms, then sufficient amounts of tissue specimens can be pooled prior to homogenization.

9.3 Difficulties with low numbers of identified proteins of pathobiochemical interest

The proteomic analysis of skeletal muscle biopsy material at the initial stages of an acquired or genetic neuromuscular disease that exhibits only relatively mild symptoms may result in a low number of identified protein candidates that exhibit differential expression patterns.

9.4 Potential solution to optimize the procedure

In general, the increase in severity of human disease is usually reflected by a rise in altered protein expression patterns. Increased symptoms are often associated with elevated numbers of significant protein hits in the pathological phenotype. Thus, if the proteomic survey of biopsy material at an early disease stage does not result in a large enough number of protein changes for proper systems biological analyses, more advanced pathological stages should be studied. The proteomic profiling of neuromuscular disorders routinely reveals changes in the expression and/or post-translational modifications in hundreds of proteoforms.

9.5 Difficulties with poorly identified lists of proteins present in tissue specimens

Depending on search parameters, the systematic mass spectrometric analysis of tissue specimens can result in a poor list of proteomic hits with an unsatisfactory coverage of peptide sequences.

9.6 Potential solution to optimize the procedure

To avoid issues with poorly identified lists of proteins present in clinical tissue samples, it is important to strictly define suitable search parameters during the data analysis steps. For the establishment of a useful multiconsensus list of distinct proteoforms of biomedical interest, the usage of critical parameters should include acceptable values for (1) peptide mass tolerance, (2) MS/MS mass tolerance, (3) the allowable number of missed peptide cleavages, (4) the minimum number of unique peptides used for protein identification, (5) a sufficient percent coverage of the total protein sequence for the unequivocal identification of distinct proteoforms, and (6) the occurrence of certain chemical modifications, including methionine oxidation and carbamido-methylation of cysteine residues.

9.7 Issues with poor data alignment using proteomic analysis software

Potential problems with poor data alignment might arise during the usage of analysis software programs, such as Progenesis QI for Proteomics.

9.8 Potential solution to optimize the procedure

Issues with data alignment can occur if the process of automatic alignment fails to properly align mass spectrometric runs during the initial round of analysis. In such a case, manual alignment vectors should be used and be included before automatic alignment is started for a second time. In general, an improvement of data alignment can be achieved by combining both manual and automatic vectors.

10. Safety considerations and standards

National legal standards and international principles in research ethics should be adhered to during the sampling, processing, storage, and analysis of clinical tissue specimens for both diagnostic and basic research purposes. Since human samples can be potentially harmful due to the presence of infectious material such as viruses or bacteria, as well as cytotoxic or radioactive material introduced to the body during diagnostic procedures or therapeutic interventions, health and safety regulations should be taken into account according to national law. For the utilization of human tissue material, such as diagnostic biopsy specimens, operational remnants or autopsy material, guidelines for anonymization of human specimens and the safe storage of patient data and research findings have to be introduced prior to starting a new project. Special safety considerations of the individual analytical steps outlined in this protocol are concerned with sodium dodecyl sulfate, dithiothreitol, and liquid nitrogen. The detergent sodium dodecyl sulfate used in the sample homogenization buffer is an irritant of exposed skin, the eyes, and the respiratory system. Therefore, during the handling of the powered form of this detergent and the preparation of the sample buffer, proper eye and face protection, a mask and protective gloves and a lab coat should be worn. The same is true for iodoacetamide, which is classified as an irritant and an acutely toxic health hazard. Special safety precautions also must be arranged when handling liquid nitrogen. Potential cryogenic burns by direct eye or skin contact should be prevented by wearing a proper face shield, protective clothing, and special cryo-gloves. Although the gaseous form of nitrogen is nontoxic, the fact that it is colorless, odorless, and tasteless and may replace oxygen in the air makes it dangerous when present at high concentration in enclosed facilities. The displacement of sufficient amounts of oxygen in the air may induce initially drowsiness and a diminished state of mental alertness, which can be followed by loss of consciousness. Hence, liquid nitrogen should never be transported in a simple Dewar container in an elevator designed for passengers or used in a nonventilated cold room. Evaporation of excess liquid nitrogen should only be carried out in a well-ventilated space or under a fume hood.

11. Alternative methods/procedures

The analytical workflow described in this protocol uses liquid nitrogen—based pulverization of tissue samples for homogenization followed by a bottom-up proteomic approach, reversed-phase liquid chromatography for peptide separation and an Orbitrap type mass spectrometer for the identification of protein species of biomedical interest. A variety of alternative approaches can be employed to analyze biopsy material. For example, depending on the amounts of available tissue material, alternative methods of homogenization can be used. If a liquid nitrogen—cooled miniature mortar and pestle system is not available for tissue grinding and pulverization, a hand-held homogenizer that is capable of handling small tissue specimens can also be used. Various commercially available types of handheld devices are suitable for the homogenization of tissue samples in the mg range. Instead of bottom-up proteomics, a more targeted top-down proteomic method can be utilized for studying clinical tissue samples. This could involve the initial protein separation using one-dimensional gel electrophoresis in combination with liquid chromatography or two-dimensional gel electrophoresis with sensitive protein dyes or the application of differential fluorescent tagging of normal control versus diseased samples. If the digestion of extracted protein populations by routine trypsination protocols, as described here, does not result in a sufficient degree of proteolytic cleavage, alternative protocols using other enzymes such as chymotrypsin, LysC, LysN, GluC, ArgC, or AspN can be used alone or in combination (LysC/trypsin; AspN/LysC/trypsin). Peptide sequencing and proteomic data analysis can also be carried out by a variety of alternative mass spectrometric methods and analysis software programs, respectively.

References

[1] Macklin A, Khan S, Kislinger T. Recent advances in mass spectrometry based clinical proteomics: applications to cancer research. Clin Proteomics 2020;17:17.

[2] Mann SP, Treit PV, Geyer PE, Omenn GS, Mann M. Ethical principles, constraints and opportunities in clinical proteomics. Mol Cell Proteomics 2021;20:100046.

[3] Dapic I, Baljeu-Neuman L, Uwugiaren N, Kers J, Goodlett DR, Corthals GL. Proteome analysis of tissues by mass spectrometry. Mass Spectrom Rev 2019;38:403—41.

[4] Nix JS, Moore SA. What every neuropathologist needs to know: the muscle biopsy. J Neuropathol Exp Neurol 2020;79:719—33.

[5] Joyce NC, Oskarsson B, Jin LW. Muscle biopsy evaluation in neuromuscular disorders. Phys Med Rehabil Clin N Am 2012;23:609—31.

[6] Staron RS, Hagerman FC, Hikida RS, Murray TF, Hostler DP, et al. Fiber type composition of the vastus lateralis muscle of young men and women. J Histochem Cytochem 2000;48:623—9.

[7] Shanely RA, Zwetsloot KA, Triplett NT, Meaney MP, Farris GE, Nieman DC. Human skeletal muscle biopsy procedures using the modified Bergström technique. J Vis Exp 2014;10:51812.

[8] Dowling P, Murphy S, Zweyer M, Raucamp M, Swandulla D, Ohlendieck K. Emerging proteomic biomarkers of X-linked muscular dystrophy. Expert Rev Mol Diagn 2019;19: 739—55.

[9]   Staunton L, Zweyer M, Swandulla D, Ohlendieck K. Mass spectrometry-based proteo-
      mic analysis of middle-aged vs. aged vastus lateralis reveals increased levels of car-
      bonic anhydrase isoform 3 in senescent human skeletal muscle. Int J Mol Med 2012;
      30:723—33.

[10] Dowling P, Gargan S, Swandulla D, Ohlendieck K. Fiber-type shifting in sarcopenia of
      old age: proteomic profiling of the contractile apparatus of skeletal muscles. Int J Mol
      Sci 2023;24(3):2415.

[11] Ohlendieck K. Two-CyDye-based 2D-DIGE analysis of aged human muscle biopsy
      specimens. Methods Mol Biol 2023;2596:265—89.

[12] Capitanio D, Moriggi M, Torretta E, Barbacini P, De Palma S, et al. Comparative pro-
      teomic analyses of Duchenne muscular dystrophy and Becker muscular dystrophy mus-
      cles: changes contributing to preserve muscle function in Becker muscular dystrophy
      patients. J Cachexia Sarcopenia Muscle 2020;11:547—63.

[13] Dowling P, Gargan S, Swandulla D, Ohlendieck K. Proteomic profiling of impaired
      excitation-contraction coupling and abnormal calcium handling in muscular
      dystrophy. Proteomics 2022;22(23—24):e2200003.

[14] Deshmukh AS, Steenberg DE, Hostrup M, Birk JB, Larsen JK, et al. Deep muscle-
      proteomic analysis of freeze-dried human muscle biopsies reveals fiber type-specific ad-
      aptations to exercise training. Nat Commun 2021;12:304.

[15] Barthelemy F, Woods JD, Nieves-Rodriguez S, Douine ED, Wang R, et al. A well-toler-
      ated core needle muscle biopsy process suitable for children and adults. Muscle Nerve
      2020;62:688—98.

[16] Dias PRF, Gandra PG, Brenzikofer R, Macedo DV. Subcellular fractionation of frozen
      skeletal muscle samples. Biochem Cell Biol 2020;98:293—8.

[17] Gargan S, Ohlendieck K. Sample preparation and protein determination for 2D-DIGE
      proteomics. Methods Mol Biol 2023;2596:325—37.

[18] Antharavally BS, Mallia KA, Rangaraj P, Haney P, Bell PA. Quantitation of proteins us-
      ing a dye-metal-based colorimetric protein assay. Anal Biochem 2009;385:342—5.

[19] Murphy S, Ohlendieck K. Protein digestion for 2D-DIGE analysis. Methods Mol Biol
      2023;2596:339—49.

[20] Dowling P, Gargan S, Zweyer M, Henry M, Meleady P, Swandulla D, et al. Protocol for
      the bottom-up proteomic analysis of mouse spleen. STAR Protoc 2020;1:100196.

[21] Antharavally BS. Removal of detergents from proteins and peptides in a spin-column
      format. Curr Protoc Protein Sci 2012. Chapter 6, Unit 6.12.

[22] Sethi S, Chourasia D, Parhar IS. Approaches for targeted proteomics and its potential
      applications in neuroscience. J Biosci 2015;40:607—27.

[23] Meola G, Bugiardini E, Cardani R. Muscle biopsy. J Neurol 2012;259:601—10.

[24] Dupree EJ, Jayathirtha M, Yorkey H, Mihasan M, Petre BA, Darie CC. A critical review
      of bottom-up proteomics: the good, the bad, and the future of this field. Proteomes 2020;
      8:E14.

[25] Kang L, Weng N, Jian W. LC-MS bioanalysis of intact proteins and peptides. Biomed
      Chromatogr 2020;34:e4633.

[26] Uzozie AC, Aebersold R. Advancing translational research and precision medicine with
      targeted proteomics. J Proteomics 2018;189:1—10.

[27] Walther TC, Mann M. Mass spectrometry-based proteomics in cell biology. J Cell Biol
      2010;190:491—500.

Sample preparation for proteomics and mass spectrometry from patient biological fluids

5

Michael Henry[1] and Paula Meleady[1,2]

[1]*National Institute for Cellular Biotechnology, Dublin City University, Glasnevin, Dublin, Ireland;*
[2]*School of Biotechnology, Dublin City University, Glasnevin, Dublin, Ireland*

1. Introduction

Biological fluids, such as serum and plasma, can be very useful for biomarker discovery as they most likely contain proteins and peptides from the site of disease and are minimally invasive during collection. They are also widely used samples for biochemical tests and biomarker measurements in the clinical setting. Serum is prepared when the blood sample is allowed to clot, centrifuged, and the blood clot is removed. Plasma is prepared through the centrifugation of blood in order to remove red and white blood cells. Anticoagulants, such as ethylenediaminetetraacetic acid (EDTA) and heparin, are used during this procedure in order to prevent blood clots from forming [1].

However, 99% of the total protein mass of plasma or serum is composed of approximately 20 proteins which is the primary disadvantage of using such specimens as they can interfere with the detection of potentially more interesting and less abundant proteins by mass spectrometry [2]. The protein concentration in serum ranges from 60 to 80 mg/mL with the concentrations of individual proteins spanning at least 10 orders of magnitude [3]. For example, albumin and IgG are two of the most abundant proteins found in serum and plasma, and contribute to approximately 80% of the total protein concentration [3]. Therefore, prior to proteomic investigation, serum and plasma samples may require a pretreatment step to remove the highly abundant proteins, allowing enrichment of the medium and lower abundant proteins which have the potential be biomarkers of disease. There is a variety of protein prefractionation techniques that can be used on serum and plasma samples [4,5]. These include lectin affinity chromatography [6], immuno-depletion [7], and an "equalization" approach which uses a combinatorial library of hexapeptide ligands coupled to beads (i.e., ProteoMiner technology) to dilute high-abundance proteins and concentrate low-abundance proteins [8].

In this chapter, we are focusing on immuno-depletion which is one of the most commonly used prefractionation techniques that can be applied to serum and plasma samples for proteomic studies. Immuno-depletion involves the removal of the most abundant proteins through an immunocapture-based technique. For example, the range of Multiple Affinity Removal System (MARS) columns (Agilent, CA, USA), can remove 1, 2, 6, 7, or 14 abundant proteins, depending on the column used. In this chapter, we describe a straightforward method to prepare a clinical patient sample that is suitable for both nongel-based LC-MS/MS and gel-based (e.g., 2D gels) proteomic approaches. We describe the use of a resin-based immuno-depletion column to remove high abundant proteins from a patient serum sample (i.e., albumin and IgG) followed by protein "In-Solution" enzymatic digestion using a nonkit-based and a simple kit-based approach.

2. Materials and equipment

- Serum sample from blood prepared and stored using standardized procedures. See Note 1.
- ProteoPrep Blue Albumin Depletion Kit (Merck, # PROTBA-1KT). See Note 2.
- Mass Spectrometry Grade Trypsin Protease (Thermo Scientific, #90057).
- ProteaseMAX Surfactant, Trypsin Enhancer (Promega, #V2071). See Note 3.
- Quick Start Bradford Protein Assay plus standards (Bio-Rad, #5000202).
- iST Sample Preparation Kit (PreOmics).
- Ammonium Bicarbonate BioUltra, ≥99.5% (Merck/Sigma−Aldrich, #09830).
- DL-Dithiothreitol (DTT) BioUltra (Merck/Sigma−Aldrich, #43815).
- Iodoacetamide BioUltra (Merck/Sigma−Aldrich, #I1149).
- Urea BioUltra (Merck/Sigma, # 51456).
- Thiourea (Merck/Sigma, #T7875).
- Acetone.
- Trifluoroacetic Acid, Optima LC/MS Grade, Fisher Chemical (Fisher Scientific, #10723857).
- Water, Optima LC/MS Grade, Fisher Chemical (Fisher Scientific, #10728098).
- Tube Rotator (e.g., SB2, Stuart).
- Microplate Spectrophotometer, e.g., MultiskanTM GO (Thermo Scientific).
- Refrigerated high speed centrifuge (e.g., Hettich Mikro 200R).
- NanoDrop One (Thermo Fisher Scientific).
- Low Protein Binding Microcentrifuge Tubes (Fisher Scientific, #90410).
- SpeedVac Vacuum dryer.

3. Before you begin

The following buffers are required for this protocol:

1. Cold (−20°C) Acetone. Cool the required volume of acetone to −20°C for 20 min before use.

2. Resuspension buffer: 6 M urea, 2 M thiourea, 10 mM Tris-HCl, pH 8 in LC-MS grade water.

3. Ammonium bicarbonate: Prepare a 50 mM and a 100 mM solution. Prepare fresh on the day of use.

4. 50 mM acetic acid solution for trypsin resuspension: Add 286 μL of glacial acetic and bring to a final volume of 10 mL in LC-MS grade water.

5. Stock MS grade trypsin solution (80X): Resuspend lyophilized trypsin using 50 mM acetic acid to a final concentration of 1 μg/μL (100 μg vial resuspended in 100 μL of 50 mM acetic acid buffer). Freeze the unused portion in 10 μL aliquots and store at −20°C.

6. ProteaseMAX surfactant and trypsin enhancer: Add 100 μL of 100 mM ammonium bicarbonate to a 100 μg vial of ProteaseMAX surfactant to give a 1% solution. Freeze in 10 μL aliquots and store at −20°C.

7. 0.5 M dithiothreitol (DTT) solution: Prepare in 50 mM ammonium bicarbonate solution. Prepare shortly before use.

8. 55 mM iodoacetamide solution: Prepare in 50 mM ammonium bicarbonate. Prepare shortly before use.

---

4. Step-by-step method details

4.1 Immuno-depletion of serum sample

1. Thaw a serum sample that has been stored at −80°C.

2. Add 0.4 mL of the ProteoPrep Blue albumin and IgG depletion resin to a spin column and centrifuge using a refrigerated centrifuge at 8000 × g for 10 s. This volume of resin is used for each serum sample.

3. Wash the resin with 0.4 mL of equilibration buffer and centrifuge at 8000 × g for 10 s.

4. Add 50 μL of the serum sample to the resin.

5. Using a tube rotator, allow the sample to incubate and rotate for 10 min at room temperature.

6. Collect the resultant eluate in the spin tube and reapply this eluate back onto the resin.

7. Allow the sample to incubate and rotate for a further 5 min at room temperature.

8. Centrifuge the spin column again at 8000 × g for 10 s and transfer the supernatant to a fresh microcentrifuge tube.

9. Wash the resin with 100 μL of equilibration buffer, centrifuge at 8000 × g for 10 s.

10. Collect this resultant eluate and pool the sample with the first collection. You should have approximately 150 μL of immuno-depleted serum in total. The samples can be aliquoted into smaller amounts and stored at −20°C until use.

11. Using the Quick Start Bradford Protein Assay (or alternative), determine the protein concentration of the immuno-depleted serum sample. Prepare a standard curve using the standards from 0 to 1 mg/mL. Aliquot 5 μL of standard and each unknown sample into individual wells of a 96 well microplate. Carry this out in triplicate. Add 250 μL of Bradford dye reagent to each sample and gently mix. Incubate at room temperature for a minimum of 5 min and a maximum of 1 h. Read the absorbance at 595 nm using a spectrophotometer, making sure to zero against the blank samples. Work out the protein concentration of each serum sample from the standard curve generated.

12. The immuno-depleted serum sample is now ready for preparation for LC-MS/MS.

4.2 Protein in-solution digestion protocol

1. If you are unsure of the composition of the buffer used in the procedure used to immuno-deplete your sample of interest, it may need processing before proteolytic digestion for mass spectrometry. This is especially important if the buffer contains detergent.

2. Precipitate the protein sample with a concentration between 50 and 200 μg with four volumes of prechilled −20°C acetone.

3. Vortex the sample and incubate in the freezer for 60 min at −20°C.

4. Pellet the protein by centrifuging the sample at 12,000 × g for 10 min at 4°C using a refrigerated microcentrifuge.

5. You will see a pellet at the bottom of the tube. Resuspend this pellet in resuspension buffer at a minimum concentration of 1.2 mg/mL.

6. Determine the protein concentration of the sample using the Quick Start Bradford Protein Assay as described in Section 3.1.

7. Aliquot 10 μg of protein to a new microcentrifuge tube making sure that the final concentration of urea in the solution is <1 M, as urea is present in the resuspension buffer used (see Note 4).

8. Add 50 mM ammonium bicarbonate to the 10 μg of protein sample to make up to a final volume of 50 μL.

9. Add 0.5 μL of 0.5 M DTT and incubate at 56°C for 20 min.

10. Add 1.35 μL of 0.55 M iodoacetamide and incubate at room temperature in the dark for 20 min. For example, you could cover the tubes with some tinfoil to keep the light away from the sample.

11. Add 1 μg of trypsin and 0.5 μL of 1% ProteaseMAX to the sample and digest at 37°C for a minimum of 3 h to overnight.

12. Centrifuge at 12,000 × g for 10 s to collect any condensate.

13. Stop the trypsin activity by the addition of TFA to a final concentration of 0.1%.

14. Determine the peptide concentration of the sample using a NanoDrop One prior to LC-MS/MS.

15. If the digest is not going to be analyzed straight away, freeze and store at −20°C.

16. Use an UltiMate 3000 nano RSLC (Thermo Scientific) (or alternative system) to perform nano-flow reverse-phased capillary high-pressure liquid chromatography (HPLC) system in combination with an Orbitrap Fusion Tribrid Mass Spectrometer (MS) (Thermo Scientific) (or alternative system) for peptide identification from the serum samples following in-solution digestion. Please refer to other chapters in the book with detailed LC-MS/MS methodologies.

4.3 Protein in-solution digestion using PreOmics iST proteomic sample preparation kit

The PreOmics iST kit is a useful in-solution proteolytic digestion sample preparation kit that is suitable for many sample types, including biological fluids such as serum. All of the reagents required for the digestion come in the kit and come preprepared.

1. The PreOmics iST kit recommends starting protein concentrations of approximately 100 µg of blood, serum, or plasma. Refer to their website and protocols for further details.
2. Add 50 µL of Lyse buffer to each sample. Place the sample in a heating block and heat at 95°C for 10 min. Allow the samples to cool to room temperature. The protein samples will be denatured, reduced, and alkylated during this step.
3. Transfer each sample to an individual labeled iST cartridge.
4. Prepare the Digest buffer and add 50 µL to each sample. The protein samples are digested for 2−3 h at 37°C.
5. Stop the enzymatic digestion by the addition of 100 µL of Stop solution and incubate for 1 min at room temperature.
6. Spin the cartridge using a centrifuge at $3800 \times g$ for 3 min.
7. Wash the peptides with 200 µL of Wash 1 solution and centrifuge the spin cartridge at $3800 \times g$ for 3 min.
8. Wash the peptides with 200 µL of Wash 2 solution and centrifuge the spin cartridge at $3800 \times g$ for 3 min.
9. Transfer the spin cartridge to a new microcentrifuge tube and add 100 µL of Eluate buffer.
10. Spin the cartridge using a centrifuge at $3800 \times g$ for 3 min.
11. Repeat this step and keep the flow through peptides from each elution.
12. Evaporate the eluted peptides using a SpeedVac and centrifuge at 48°C until complete dryness.
13. Dried peptides are either resuspended in 50 µL of LC-Load solution for direct LC-MS analysis or can be frozen and stored at −20°C until use.
14. If direct LC-MS analysis is being carried out, determine the peptide concentration using a NanoDrop One.
15. If the peptide samples require desalting and sample concentration, use a 10 µL Zip Tip (0.6 µL bed volume).

16. Use an UltiMate 3000 nano RSLC (Thermo Scientific) (or alternative system) to perform nanoflow reverse-phased capillary high-pressure liquid chromatography (HPLC) system in combination with an Orbitrap Fusion Tribrid Mass Spectrometer (MS) (Thermo Scientific) (or alternative system) for peptide identification from the serum samples following in-solution digestion. Please refer to other chapters in the book with detailed LC-MS/MS methodologies.

5. Anticipated results

The immuno-depletion method described in this chapter is a straightforward methodology for removal of high abundant proteins (e.g., IgG and serum albumin) from a biological fluid, i.e., serum from patient blood. In our laboratory we have used various immuno-depletion strategies for removal of high abundant proteins from different types of biological fluid for proteomic analysis using LC-MS/MS. For example, Fig. 5.1 shows vitreous fluid from the eye of a patient with uveal melanoma before and after immuno-depletion using the Multiple Affinity Removal Spin

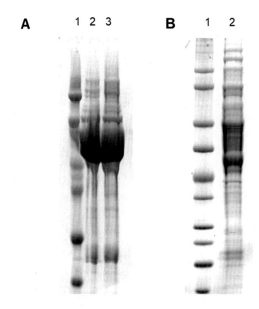

FIGURE 5.1

Vitreous fluid from the eye separated using SDS-PAGE showing a decrease in high abundant proteins following immuno-depletion. (A) Nondepleted vitreous fluid; lane 1—molecular weight marker, lanes 2 and 3—20 μL of nondepleted vitreous fluid. (B) Immuno-depleted vitreous fluid using the MARS 14 spin column; lane 1—molecular weight markers, lane 2—20 μL of immuno-depleted vitreous fluid.

(MARS) Cartridge Human 14 spin column that can be purchased from Agilent. This column removes 14 of the most abundant proteins found in human serum/plasma. The images show the successful depletion of high abundant proteins in the vitreous fluid sample, and an enrichment of lower abundant proteins in the sample.

6. Optimization and troubleshooting/notes

1. Note 1: Serum and plasma should be prepared from blood according to standardised procedures [9]. For example, after collection of a blood sample, allow the blood to clot by leaving it undisturbed at room temperature for approximately 15–30 min. The clot can then be removed by centrifuging at 1000–2000 × g for 10 min in a refrigerated centrifuge. The resultant supernatant is the serum fraction and should be stored at −80°C in aliquots (to reduce the number of freeze/thaw cycles) if not used immediately. Plasma can be collected by using commercially available anticoagulant-treated tubes, e.g., EDTA-treated, citrate-treated. The cells can be removed from the plasma by centrifugation for 10 min at 1000–2000 × g using a refrigerated centrifuge. Platelets in the plasma sample can be depleted by centrifugation at 2000 × g for 15 min. The plasma fraction should be stored in aliquots at −80°C if not used immediately.

2. Note 2: There are many commercially available immuno-depletion kits and columns that can remove high abundant proteins in serum such as IgG, serum albumin, etc. Examples include ProteoPrep (SIGMA/Merck), ProteoSpin (NORGEN BioTek Corp), Proteome Purify (R&D Systems/Bio-techne), High Select Depletion Spin Columns (Thermo Scientific), and Multiple Affinity Removal (MARS) range (Agilent). These kits and columns are not just confined to use on serum but can be used for immuno-depletion of other biological fluids (e.g., plasma, vitreous fluid, wound exudate, etc.).

3. Note 3: We use ProteaseMax Surfactant Trypsin Enhancer as when we carry out protein in-solution digestions we find protein solubilization is improved.

4. Note 4: Trypsin activity is inhibited by urea at concentrations >1 M.

7. Safety considerations

The techniques described here use biological materials and chemicals. Personal protection equipment must be worn at all times. Working with chemicals one should be very familiar with their Material Safety Data Sheets ahead of use. Validated biological safety cabinets and Fume hoods should only be used. All waste generated (biological and chemical) should be disposed of in accordance with local guidelines and procedures.

8. Summary

In this chapter, we have provided a methodology for the immuno-depletion of high abundant proteins from serum (e.g., albumin and IgGs). The protocol can be easily adapted to various types of biological fluids and the types of columns/resins being used. Vendor websites have a lot of information on the different types of resins/columns available and details on the various high abundant proteins that they deplete.

References

[1] Lam NYL, Rainer TH, Chiu RWK, Lo YMD. EDTA is a better anticoagulant than heparin or citrate for delayed blood processing for plasma DNA analysis. Clin Chem 2004; 50:256−7. https://doi.org/10.1373/clinchem.2003.026013.

[2] Fernández-Costa C, Calamia V, Fernández-Puente P, Capelo-Martínez J-L, Ruiz-Romero C, Blanco FJ. Sequential depletion of human serum for the search of osteoarthritis biomarkers. Proteome Sci 2012;10:55. https://doi.org/10.1186/1477-5956-10-55.

[3] Adkins JN, Varnum SM, Auberry KJ, Moore RJ, Angell NH, et al. Toward a human blood serum proteome. Mol Cell Proteomics 2002;1:947−55. https://doi.org/10.1074/mcp.M200066-MCP200.

[4] Kaur G, Poljak A, Ali SA, Zhong L, Raftery MJ, Sachdev P. Extending the depth of human plasma proteome coverage using simple fractionation techniques. J Proteome Res 2021;20:1261−79. https://doi.org/10.1021/acs.jproteome.0c00670.

[5] Palstrøm NB, Rasmussen LM, Beck HC. Affinity capture enrichment versus affinity depletion: a comparison of strategies for increasing coverage of low-abundant human plasma proteins. IJMS 2020;21:5903. https://doi.org/10.3390/ijms21165903.

[6] Dahabiyeh LA, Tooth D, Barrett DA. Profiling of 54 plasma glycoproteins by label-free targeted LC-MS/MS. Anal Biochem 2019;567:72−81. https://doi.org/10.1016/j.ab.2018.12.011.

[7] Xu S, Jiang J, Zhang Y, Chen T, Zhu M, Fang C, et al. Discovery of potential plasma protein biomarkers for acute myocardial infarction via proteomics. J Thorac Dis 2019;11: 3962−72. https://doi.org/10.21037/jtd.2019.08.100.

[8] Wang X, Chen M, Dai L, Tan C, Hu L, et al. Potential biomarkers for inherited thrombocytopenia 2 identified by plasma proteomics. Platelets 2022;33:443−50. https://doi.org/10.1080/09537104.2021.1937594.

[9] Tuck MK, Chan DW, Chia D, Godwin AK, Grizzle WE, et al. Standard operating procedures for serum and plasma collection: early detection research network consensus statement standard operating procedure integration working group. J Proteome Res 2009;8:113−7. https://doi.org/10.1021/pr800545q.

Preparation of bacterial and fungal samples for proteomic analysis

6

Magdalena Piatek and Kevin Kavanagh

Department of Biology, SSPC, the Science Foundation Ireland Research Centre for Pharmaceuticals, Maynooth University, Maynooth, Co. Kildare, Ireland

Abbreviations

FDR	False discovery rate
GO	Gene ontology
GOBP	Gene ontology biological process
GOCC	Gene ontology cellular component
GOMF	Molecular function
KEGG	Kyoto Encyclopedia of Genes and Genomes
LFQ	Label-free quantitative
PBS	Phosphate buffered saline
PMSF	Phenylmethylsulfonyl fluoride
SSDA	Statistically significant and differentially abundant
TFA	Trifluoroacetic acid
TLCK	Tosyllysine chloromethyl ketone hydrochloride

1. Section 1 (bacteria)

1.1 Introduction

Proteomic techniques can deeply enhance our understanding of the workings of pathogenic and beneficial microorganisms and host—microbe interactions at a molecular level. Bacteria are attractive organisms to study due to less complex, prokaryotic proteomes, simple culture conditions, and rapid cell turnover for fast generation of results [1,2]. Bacterial infections continue to pose a threat to human health worldwide as one of the leading causes of death [3]. The rising incidence of antimicrobial drug resistance demands discovery of novel and/or repurposed therapies and proteomics plays an important role in shedding light on the response to these antimicrobial agents, adaptation to the environment, antibiotic resistance mechanisms and virulence [4].

This section will discuss sample preparation and data analysis for two bacterial species: *Pseudomonas aeruginosa* (Gram-negative) and *Staphylococcus aureus* (Gram-positive), although the protocol may be used for a diverse range of species,

Proteomics Mass Spectrometry Methods. https://doi.org/10.1016/B978-0-323-90395-0.00005-X

with the exception of adjustments in protein extraction methods depending on growth requirements and physical structure.

The ESKAPE pathogens (*Enterococcus faecium, Staphylococcus aureus, Klebsiella pneumoniae, Acinetobacter baumannii, Pseudomonas aeruginosa,* and *Enterobacter* species) are the most frequent cause of nosocomial infection and have rapidly emerged as drug-resistant isolates [5]. *P. aeruginosa* (Fig. 6.1) has the capacity to cause infection in both immunocompromised and immunocompetent individuals [7]. The rather large genome ('6 Mbp) permits expression of a repertoire of genes used for adaptation to hostile environments, virulence traits, and multidrug resistance [8,9]. This pathogen has become extremely challenging to treat particularly in the case of neutropenic, cancer, AIDS, transplant and cystic fibrosis patients with increased morbidity and mortality rates among patients [10].

S. aureus (Fig. 6.2) resides as a commensal of the skin microflora offering protection against harmful pathogens [12]. Changes/disruption in the environment, host immunity, and the organism's virulence capabilities can promote cutaneous infection [13]. Mutations in the penicillin-binding protein target site have given rise to multi-drug resistant strains including the leading cause of nosocomial infection—Methicillin-resistant *S. aureus* (MRSA), which has rendered an entire class of antibiotics (Beta-lactams) ineffective [14].

Traditional gel-based techniques lack sensitivity and prevent studies of whole proteomes [15]. The evolution of gel-free mass spectrometry–based techniques have coincided with improved diagnostic and therapeutic strategies in the fight against bacterial pathogens [16].

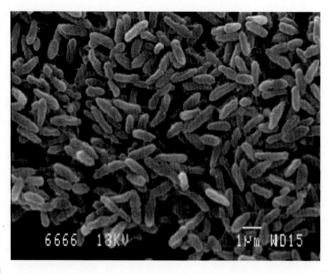

FIGURE 6.1

Microscopy image of *Pseudomonas aeruginosa* biofilm [6].

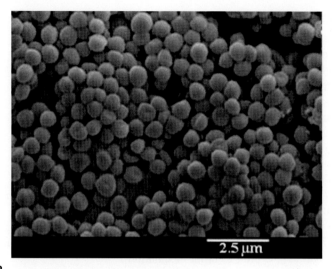

FIGURE 6.2

Staphylococcus aureus microscopy image [11].

1.2 Before you begin

1.2.1 Protein extraction

1. Nutrient media for bacterial growth (e.g., nutrient broth (Oxoid)).
2. Lysis buffer: 6 M urea, 2 M thiourea, 0.1 M Tris-hydrochloride, dissolved in deionized water, adjusted to pH 8.0.
3. Protease inhibitors: Aprotinin, Tosyllysine Chloromethyl Ketone hydrochloride (TLCK), leupeptin, pepstatin A stocks (1 mg/mL), and phenylmethylsulfonyl fluoride (PMSF; 100 mM) (alternatively can use a protease inhibitor cocktail tablet (Complete Series, Roche)).
4. Phosphate buffered saline (PBS).
5. Acetone (99+%; ice cold).
6. Spectrophotometer capable of measuring absorbance (A) at 600 nm.

1.2.2 In solution protein digestion

1. Sample resuspension buffer: 6 M urea, 2 M thiourea, 0.1 M Tris-HCl, pH 8.0.
2. 50 mM ammonium bicarbonate: Prepared fresh in deionized water.
3. 0.5 M dithiothreitol (DTT): Prepare fresh in 50 mM ammonium bicarbonate.
4. 0.5 M iodoacetamide (IAA): Prepare just before use in 50 mM ammonium bicarbonate and prevent light exposure.
5. ProteaseMAX Surfactant Trypsin Enhancer (MyBio V2071) (stock (1% (w/v)): resuspend 1 mg in 100 μL of 50 mM ammonium bicarbonate immediately before use.
6. Trypsin, sequencing grade (MyBio V5111; 0.5 μg/μL): resuspend sequencing grade trypsin vial (V5111 Promega) with trypsin resuspension buffer.

1.2.3 Peptide purification

1. C-18 spin columns (Pierce).
2. Trifluoroacetic acid (TFA; ≥99%).
3. Sample buffer: 2% TFA in 20% acetonitrile (ACN).
4. Activation solution: 50% ACN and 50% deionized water.
5. Equilibration buffer: 0.5% TFA in 5% ACN.
6. Wash buffer: same as equilibration buffer.
7. Elution buffer: 70% ACN in deionized water.
8. Loading buffer for QExactive: 0.05% TFA in 2% ACN.

1.2.4 Bacteria details

Bacteria: *Pseudomonas aeruginosa* (Gram-negative), *Staphylococcus aureus* (Gram-positive).

Growth conditions: Liquid cultures are grown in desired media (e.g., nutrient broth) at 37°C in an orbital shaker at 200 rpm. (Bacterial stocks are maintained on nutrient agar at 4°C prior to broth inoculation.)

1.3 Step-by-step method details

1.3.1 Protein extraction—day 1

1. Grow cells to stationary phase.
2. Inoculate fresh, sterile media (5–50 mL) with aliquots of 18 h culture containing the desired concentration of cells and allow cells to grow at 37°C until mid-exponential phase. (It is recommended to use a minimum of four replicates per sample group.)
3. Pellet cells by centrifuging at $2000 \times g$ for 15 min.
4. Wash cells in sterile PBS twice.
5. Resuspend cell pellet with 1 mL lysis buffer containing protease inhibitors (add 10 μL of each protease inhibitor per 1 mL of lysis buffer to obtain final protease inhibitor concentrations of 10 μg/mL and 1 mM/mL for PMSF) and keep chilled on ice. Sonicate resuspended cells with a sonicator probe for 10 s three times at 50% power.
6. Pellet cell debris via centrifugation at $14{,}500 \times g$ for 8 min.
7. Retain the supernatant and use to quantify protein concentration via the Bradford protein assay.

1.3.2 Bradford protein assay (see Note 1)

1. To quantify protein concentrations, prepare protein assay dye reagent (Bio-rad) stock solution in deionized water in a 1:4 ratio, biorad:water.
2. To measure protein concentrations in the sample, add supernatant to Biorad stock solution in 1 in 50 ratio, sample to stock.
3. Aspirate samples in a cuvette and read on a spectrophotometer set to the "Bradford" setting.

4. Standardize protein concentrations to obtain 75–100 μg in ice-cold acetone in a 1:3 ratio, sample:acetone.
5. Leave samples to precipitate in acetone overnight at −20°C.

1.3.3 In solution protein digestion—day 2

1. Centrifuge samples at 13,000 × g for 10 min at 4°C.
2. Discard acetone and allow any remaining acetone to evaporate without over-drying the protein pellet.
3. Resuspend samples in 25 μL of sample resuspension buffer (samples may be sonicated in a water bath briefly to aid peptide resuspension). Remove an aliquot (e.g., 5 μL) from the sample and *set aside.
4. To the remainder of the sample, add 105 μL of 50 mM ammonium bicarbonate.
5. Reduce the protein with DTT (1 μL) for 20 min at 56°C.
6. Allow the samples to cool to room temperature and alkylate with IAA (2.7 μL) for 15 min in the dark.
7. Digest the protein with sequencing grade trypsin at a trypsin:protein ration of 1: 40 (Promega, Ireland). This is added with 1 μL of ProteaseMAX and incubated for 18 h at 37°C.

Notes

Note 1: Sample aliquots can be used to quantify protein prior to mass spectrometry loading. This can be carried out using the Qubit Protein Assay Kit (Invitrogen) as per manual instructions. With the remaining sample, 1-D SDS-PAGE ensures an additional level of quality control prior to in-solution digestion.

1.3.4 Peptide purification—day 3 (see Note 2)

1. Inhibit digestion with TFA (1 μL) and incubate at room temperature for 5 min.
2. Centrifuge the samples at 13,000 × g for 10 min.
3. Combine the resulting supernatants with sample buffer in 1:3 ratio, sample buffer:sample. Apply activation solution (200 μL) to each C-18 column placed in collection tubes.
4. Centrifuge the sample tubes at 1500 × g for 1 min and discard the flow through. Repeat this step.
5. Add equilibration solution (200 μL) in the same manner.
   Notes
   Note 2: C-18 spin columns hold up to 30 μg of protein, therefore high protein yields (determined via Qubit quantification) will need to be fractioned accordingly prior to C-18 column loading.
6. Samples (10–150 μL) are applied to the top of the resin bed and columns are placed in new receiver tubes.
7. Centrifuge the samples at 1500 × g for 1 min.
8. Recover the flow through and reapply to the resin.
9. Centrifuge and repeat this step once more to ensure complete peptide binding.

10. Wash the samples with wash buffer (200 µL) three times to eliminate contaminants. Place columns in new receiver tubes and elute peptides with 25 µL of elution buffer three times.
11. Dry the resulting eluant (75 µL) in a SpeedVac concentrator for 2 h at 39°C. Resuspend the samples in QExactive loading buffer to give a final peptide concentration between 500 and 1000 ng/µL in all samples.
12. Sonicate the samples in a sonication bath for 5 min to aid resuspension and centrifuge at $14,500 \times g$ for 5 min.
13. Transfer the supernatant into LC-MS/MS vials for QExactive loading.

2. Section 2 (fungi)

2.1 Introduction

Fungi colonize many diverse habitats and therefore must compete against their microbial competitors in the form of chemical defense [17]. The resulting production of secondary metabolites has led to the isolation of therapeutic agents with anticancer, immunosuppressive, hypercholesterolemic, and antimicrobial properties [17–19]. The fungal primary metabolism has also been exploited for the production of an abundance of agricultural, cosmetic, and industrial commodities [17,20].

As eukaryotes, fungi are attractive model organisms for the study of cellular processes in mammals. Almost a quarter of yeast genes are homologous to humans and so have been employed for genetic, developmental, and pathological studies with easier, faster, and cheaper cultivation and generation of results [21].

By contrast, fungi are responsible for an estimated 1.6 million deaths due to infections annually, a number likely to increase with continued use of antibiotics and immunosuppressive therapies, climate change and the emergence of drug-resistant isolates [22,23]. Immunocompromised individuals are particularly susceptible to infections that can range from (recalcitrant) superficial infections to life-threatening systemic infections [24]. Although *Candida* species are more frequently isolated in ICU patients, mortality rates are exacerbated by the filamentous fungus, *Aspergillus fumigatus* (Fig. 6.3), which according to the CDC, exceeds 50% in invasive aspergillosis patients [26,27]. Similarities between fungi and host cells have hindered antifungal drug development due to limitations in drug selectivity [28]. Consequently, there are limited treatment options to combat the increasing demand for novel and/or repurposed therapies. Additionally, phytopathogenic fungi pose a serious threat to crop productivity resulting in an estimated global loss of 200 billion dollars yearly [29].

Fungi are fundamental components of our ecosystems. Advancements in proteomic studies open avenues for more extensive mycological research to enhance our understanding in treating harmful pathogens and utilizing these greatly valuable organisms.

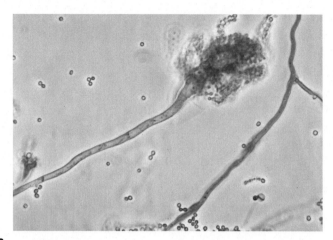

FIGURE 6.3

Aspergillus fumigatus conidiophore and individual conidia [25].

2.2 Before you begin—yeast

This protocol is suitable for use with yeast species. In the case described here, the protocol for extracting protein from the pathogenic yeast, *Candida albicans,* will be described.

Growth media: YEPD (10–50 mL): 2% (w/v) glucose (Sigma), 2% (w/v) peptone (Sigma–Aldrich), 1% (w/v) yeast extract (Fisher BioReagents). To make YEPD agar plates add 2% (w/v) agar to above. All other materials are as previously mentioned in Section 1.

Growth conditions: Inoculate 50 mL of YEPD broth with *Candida albicans* culture from YEPD agar plate. Incubate at 30 or 37°C at 200 rpm in an orbital shaker for 18 h to the stationary phase. (*Candida* stocks can be maintained on YEPD agar plates at 4°C.)

2.3 Step-by-step method details—yeast

2.3.1 Protein extraction—day 1

1. Grow cells for 18 h in YEPD broth to the stationary phase (approximately 1×10^8/mL).
2. Remove a sample of culture, dilute 1/100 with PBS, and enumerate using a hemocytometer to add the desired number of cells to fresh, sterile media and grow for the desired period of time, ensuring an appropriate cell density for protein extraction.
3. Harvest the cells by centrifugation at $2000 \times g$.
4. Wash cells twice with sterile PBS.

5. Resuspend the final pellet in lysis buffer containing a range of protease inhibitors (see Section 1 protease inhibitor list). Lyse cells either via a sonication probe or bashing beads and follow methods as per Section 1 for protein extraction, in-solution digestion and sample purification.

2.4 Before you begin—filamentous fungi

This protocol will describe the steps used to extract proteins from *Aspergillus fumigatus* ATCC 26933.

Growth Media (e.g., Sabouraud dextrose liquid medium (Oxoid) or Czapek Dox).

Growth conditions: *A. fumigatus* is grown on Sabouraud dextrose agar at 37°C for 5 days.

2.5 Step-by-step method details—filamentous fungi

1. Harvest *A. fumigatus* conidia from a plate with 10 mL 0.01% Tween-20 in PBS and wash twice with PBS.
2. Enumerate conidia via hemocytometry and add to fresh media at the desired concentration (e.g., 5×10^5 conidia/mL).
3. Grow cultures for 24–72 h at 37°C in an orbital shaker at 200 rpm. Harvest mycelium by straining through a double layer of Mira cloth and record the wet weight of the sample.
4. Grind the hyphal mass with liquid nitrogen to a fine powder using a mortar and pestle.
5. Add lysis buffer containing protease inhibitors (see Section 1), 4 mL/g hyphae. Disrupt the suspension using a sonication probe (Bandelin Sonopuls, Bandelin electronic, Berlin) at 50% power, three times for 10 s each and cooling the sample on ice between each sonication.
6. Centrifuge the cell lysate at $14,500 \times g$ at 4°C for 10 min to pellet the cell debris.
7. Quantify the protein using the Bradford method, and acetone precipitate the protein samples as per Section 1.
8. Day 2 and 3 in-solution digestion and sample purification are carried out as per Section 1.

3. Section 3 label-free mass spectrometry
3.1 Step-by-step method details
3.1.1 LC-MS/MS analysis

1. Biological replicates from all sample groups (resuspended in QExactive loading buffer (see Section 1—Day 3 peptide purification) are loaded onto a QExactive (ThermoFisher Scientific) high-resolution accurate mass spectrometer connected to a Dionex Ultimate 3000 (RSLCnano) chromatography system.

2. The instrument is commanded by LC/MS Xcalibur software and the parameters are universal for all organism types discussed in Sections 1 and 2.
3. A 50 cm EASY-spray PepMap C18 column with a 75 μm diameter (2 μm particle size) separates peptides by an increasing 3%−90% acetonitrile gradient.
4. A 133 min reverse-phase gradient at a flow rate of 300 nL/min is used.
5. All data are obtained with the MS device running in an automatic-dependent switching mode.
6. A full MS scan is set at 70,000 resolution with a scan range of 400−1600 *m/z* proceeded by an MS/MS scan at 17,500 resolution and range of 200−2000 *m/z*.
7. The 15 most intense ions prior to MS/MS are selected.

3.1.2 Statistical analysis of LC-MS data

1. Protein identification from the MS/MS data is performed using the Andromeda search engine in MaxQuant (www.maxquant.org) using the latest version to correlate the data against an annotated database retrieved from Uniprot SWISS-PROT for the desired organism.
2. For example, the *P. aeruginosa* PAO1 reference proteome (strain ATCC 15692/DSM 22644/CIP104116/JCM 14847/LMG 12228/1C/PRS 101/PAO1), with a protein count of 5564; accessed in 2022, can be downloaded from Uniprot.
3. Maxquant search parameters are set to Orbitrap as the instrument, and label-free quantification is selected.
4. The first search peptide tolerance is set to 20 and 4.5 ppm as the main search peptide tolerance with cysteine carbamidomethylation set as a fixed modification and methionine oxidation and protein N-acetylation as variable modifications.
5. A maximum of two missed cleavage sites is permitted.
6. The peptide and protein false discovery rates (FDR) are set to 1%.
7. Only peptides with a minimum length of seven amino acids are identified. Results processing, statistical analyses and graphics generation can be conducted using Perseus (www.maxquant.org) whereby protein abundances are based on normalized label-free quantification (LFQ) intensity values.
8. All initially identified proteins are filtered to remove contaminants and peptides identified by site and the resulting matrix is $\log_2$-transformed.
9. All biological replicates are designated into their corresponding sample group (e.g., "Control" or "Treatment").
10. Proteins unidentified in at least three out of four replicates are removed from the dataset.
11. The data are imputed to replace missing values with random values extracted from a normal distribution with a downshift 1.8 times the mean standard deviation of all measured values and a width 0.3 times this standard deviation. Normalized LFQ intensity values can be used to construct a principal component analysis (Fig. 6.4).
12. To determine the number of proteins identified in each sample group and those which are exclusively expressed, a numeric Venn diagram can be conducted and used for subsequent analysis.

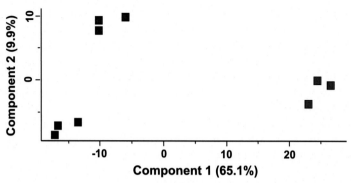

FIGURE 6.4

Principal component analysis of three various sample groups comprising of three biological replicates in each (as shown highlighted *red, blue,* and *black*). Samples are clustered together based on similarities to one another and in this example, the combined variance amounts to 75%.

13. Gene ontology (GO) mapping via the UniProt gene ID is used to query the annotation file which can be downloaded through Perseus for the desired organism to acquire terms for gene ontology biological process (GOBP), gene ontology cellular component (GOCC), molecular function (GOMF), and Kyoto Encyclopedia of Genes and Genomes (KEGG).

14. To generate volcano plots, $\log_2$ fold changes are plotted on the *x*-axis and $-\log P$ values on the *y*-axis for each comparison. Various pathways and processes are listed in the "Categories" function which can be highlighted to visualize distribution on the corresponding volcano plot (Fig. 6.5).

15. Statistically significant (ANOVA, $P < .05$) and differentially abundant proteins (with the desired fold change, i.e., a minimum of 1.5) can be retrieved from this dataset.

16. $\log_2$ transformed LFQ intensity values are subjected to *z*-score normalization and hierarchal clustering of expression SSDA values using Euclidean distance and average linkage.

17. For further GO and KEGG enrichment analysis, a Fisher's exact test (Benjamini-Hochberg, FDR cut-off of 5%) is selected for enrichment using UniProt key words, GOBP, GOCC, GOMF and KEGG (Fig. 6.6).

4. Data availability

For manuscript publication, MS proteomics data and MaxQuant output files can be uploaded to the ProteomeXchange Consortium via the PRIDE partner repository. Following successful file upload, a dataset identifier is provided to permit public data sharing [30].

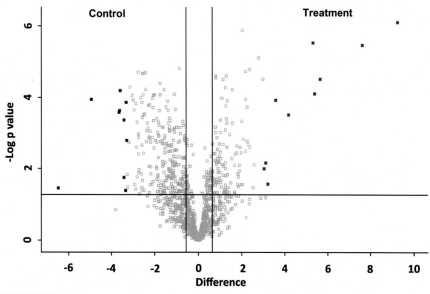

FIGURE 6.5

Volcano plot representing the distribution of all identified proteins postfiltration of contaminants generated from a two-sample *t*-test (*P* < .05). Statistically significant (*P* < .05) proteins are positioned above the horizontal *black* line and differentially abundant proteins (fold change ≥1.5) are shown to the right and left of the vertical *black* lines. The top 10 most increased abundance proteins are highlighted in *red*, and those decreased are shown in *blue*. Protein names/IDs can be retrieved from Perseus and annotated.

Cluster	Function	Proteins in cluster
A	Carbohydrate metabolism	30
	Pathogenesis	22
	Glycolysis/gluconeogenesis	18
B	RNA metabolic process	33
	RNA biosynthetic process	20
	Transcription	16

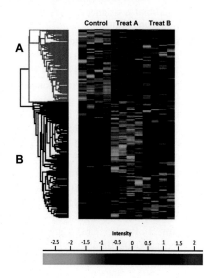

FIGURE 6.6

Hierarchal clustering of statistically significant and differentially abundant (SSDA) proteins between three sample groups. Two distinct clusters, A and B, are resolved which include proteins of similar expression profiles. Statistical enrichment analysis revealed gene ontology (GO) and KEGG terms, some of which are outlined in the table.

5. Concluding remarks

Advances in proteomic analysis have revolutionized our knowledge of how microbes interact with the host and how they respond to antimicrobial agents. This knowledge can be used to better understand host—pathogen interactions and opens the way to the development of novel antimicrobial agents. Microbial studies are just one of many applications of proteomics—a powerful tool in human health research which has provided invaluable insight into novel biomarkers, disease prognosis and progression. Despite this, there is scope for further improvements. Complex organisms and/or physiological processes, vast data sets and intricate and costly instrumentation requiring extensive skills and knowledge are among several limitations of proteomic studies.

References

[1] Stekhoven DJ, et al. Proteome-wide identification of predominant subcellular protein localizations in a bacterial model organism. J Proteonomics 2014;99. https://doi.org/10.1016/j.jprot.2014.01.015.

[2] Meunier A, Cornet F, Campos M. Bacterial cell proliferation: from molecules to cells. FEMS Microbiol Rev 2021;45(1):fuaa046. https://doi.org/10.1093/femsre/fuaa046.

[3] Thompson T. The staggering death toll of drug-resistant bacteria. Nature 2022. https://doi.org/10.1038/d41586-022-00228-x.

[4] Tsakou F, et al. The role of proteomics in bacterial response to antibiotics. Pharmaceuticals (Basel) 2020;13(9):214. https://doi.org/10.3390/ph13090214.

[5] Mulani MS, et al. Emerging strategies to combat ESKAPE pathogens in the era of antimicrobial resistance: a review. Front Microbiol 2019;10:539. https://doi.org/10.3389/fmicb.2019.00539.

[6] Deligianni E, et al. Pseudomonas aeruginosa cystic fibrosis isolates of similar RAPD genotype exhibit diversity in biofilm forming ability in vitro. BMC Microbiol 2010;10. https://doi.org/10.1186/1471-2180-10-38.

[7] Migiyama Y, et al. Pseudomonas aeruginosa bacteremia among immunocompetent and immunocompromised patients: relation to initialantibiotic therapy and survival. Jpn J Infect Dis 2016;69(2). https://doi.org/10.7883/yoken.JJID.2014.573.

[8] Silby MW, et al. Pseudomonas genomes: diverse and adaptable. FEMS (Fed Eur Microbiol Soc) Microbiol Rev 2011;35(4). https://doi.org/10.1111/j.1574-6976.2011.00269.x.

[9] Dubern JF, et al. Integrated whole-genome screening for Pseudomonas aeruginosa virulence genes using multiple disease models reveals that pathogenicity is host specific. Environ Microbiol 2015;17(11). https://doi.org/10.1111/1462-2920.12863.

[10] Sadikot RT, et al. Pathogen—host interactions in pseudomonas aeruginosa pneumonia. Am J Respir Crit Care Med 2005;171(11):1209—23. https://doi.org/10.1164/rccm.200408-1044SO.

[11] Khelissa SO, et al. Bacterial contamination and biofilm formation on abiotic surfaces and strategies to overcome their persistence. J Mater Environ Sci 2017;8(9).

[12] Jenkins A, et al. Differential expression and roles of Staphylococcus aureus virulence determinants during colonization and disease. mBio 2015;6(1). https://doi.org/10.1128/mBio.02272-14.

[13] Creech CB, Al-Zubeidi DN, Fritz SA. Prevention of recurrent staphylococcal skin infections. Infect Dis Clin North Am 2015;29(3):429—64. https://doi.org/10.1016/j.idc.2015.05.007.

[14] Katayama Y, Zhang HZ, Chambers HF. PBP 2a Mutations producing very-high-level resistance to beta-lactams. Antimicrob Agents Chemother 2004;48(2). https://doi.org/10.1128/AAC.48.2.453-459.2004.

[15] Baggerman G, et al. Gel-based versus gel-free proteomics: a review. Comb Chem High Throughput Screen 2005;8(8). https://doi.org/10.2174/138620705774962490.

[16] Kathera C. Microbial proteomics and their importance in medical microbiology. Recent Dev Appl Microbiol Biochem 2018:21—31. https://doi.org/10.1016/B978-0-12-816328-3.00003-9.

[17] Hyde KD, et al. The amazing potential of fungi: 50 ways we can exploit fungi industrially. Fungal Divers 2019;97. https://doi.org/10.1007/s13225-019-00430-9.

[18] Manzoni M, Rollini M. Biosynthesis and biotechnological production of statins by filamentous fungi and application of these cholesterol-lowering drugs. Appl Microbiol Biotechnol 2002;58:555—64. https://doi.org/10.1007/s00253-002-0932-9.

[19] Künzler M. How fungi defend themselves against microbial competitors and animal predators. PLoS Pathog 2018;14(9):e1007184. https://doi.org/10.1371/journal.ppat.1007184.

[20] Dowd CJ, Kelley B. Purification process design and the influence of product and technology platforms. In: Comprehensive biotechnology. 2nd ed. 2011. https://doi.org/10.1016/B978-0-08-088504-9.00137-9.

[21] Liu W, et al. From saccharomyces cerevisiae to human: the important gene co-expression modules. Biomed Rep 2017;7(2). https://doi.org/10.3892/br.2017.941.

[22] Low CY, Rotstein C. Emerging fungal infections in immunocompromised patients. F1000 Med Rep 2011;3(1). https://doi.org/10.3410/M3-14.

[23] Fausto A, Rodrigues ML, Coelho C. The still underestimated problem of fungal diseases Worldwide. Review Front Microbiol 2019;10:214. https://doi.org/10.3389/fmicb.2019.00214.

[24] Odom RB. Common superficial fungal infections in immunosuppressed patients. J Am Acad Dermatol 1994;31(3). https://doi.org/10.1016/S0190-9622(08)81269-8.

[25] Bandres M, Modi P, Sharma S. [Figure, Aspergillus fumigatus. Contributed by centers for disease control and prevention (public domain)]. StatPearls — NCBI bookshelf. Ncbi.nlm.nih.gov; 2022. Available from: https://www.ncbi.nlm.nih.gov/books/NBK482464/figure/article-17893.image.f1/ [Accessed 14 September 2022].

[26] Turner SA, Butler G. The Candida pathogenic species complex. Cold Spring Harb Perspect Med 2014;4(9). https://doi.org/10.1101/cshperspect.a019778.

[27] Beer KD, et al. Multidrug-resistant aspergillus fumigatus carrying mutations linked to environmental fungicide exposure — three states, 2010—2017. MMWR Morb Mortal Wkly Rep 2018;67(38). https://doi.org/10.15585/mmwr.mm6738a5.

[28] Mazu TK, et al. The mechanistic targets of antifungal agents: an overview. Mini-Rev Med Chem 2016;16(7). https://doi.org/10.2174/1389557516666160118112103.

[29] Villamizar-Gallardo R, Osma JF, Ortíz-Rodriguez OO. Regional evaluation of fungal pathogen incidence in Colombian cocoa crops. Agriculture 2019;9(3). https://doi.org/10.3390/agriculture9030044.

[30] Perez-Riverol Y, et al. The PRIDE database and related tools and resources in 2019: improving support for quantification data. Nucleic Acids Res 2019;47(D1):D442—50. https://doi.org/10.1093/nar/gky1106.

Urinary extracellular vesicles isolated by hydrostatic filtration dialysis without Tamm—Horsfall protein interference for Mass Spectrometry analysis

Luca Musante

School of Veterinary Medicine, University of Pennsylvania, Philadelphia, PA, United States

1. Introduction

Urinary Extracellular Vesicles (uEVs) are made of different types of lipid membrane—enclosed particles of various sizes (40—600 nm) and morphology [1,2]. All the cells lining the urogenital apparatus actively secrete uEVs into the lumen of the nephron, the urinary tract (ureter, bladder, and urethra), and genital glands in males (prostate and seminal glands).

Since the first thorough characterization of exosomes in urine, the protein composition analysis as determined by Mass Spectrometry (MS) identified more than 1000 proteins, many of which are related to well-known kidney diseases [3,4]. Since then, exosomes and other types of uEVs have shown their potential as source of biomarkers that allow a precise localization of the cell of origin. Thus, uEVs faithfully reflect pathophysiologic changes in the kidney [5], bladder [6], and prostate [7], therefore reinforcing their utility as a liquid biopsy to monitor kidney and urogenital functions.

uEVs are particularly suitable for biomarker discovery as a result of their stability in urine [8], resistance to extreme ionic solutions [9], and harsh treatments during the isolation protocol [10—12]. However, uEV research is extremely challenging due to the inherent complex composition of vesicle types and their broad heterogeneity in size and morphology. Many protocols for uEV isolation have been proposed. The majority require long and labor-intensive steps to reach good purity of uEVs [13]. The main hurdle is represented by Tamm Horsfall Protein (THP) or uromodulin, the most abundant protein in the urine, which interferes with a successful MS analysis.

Proteomics Mass Spectrometry Methods. https://doi.org/10.1016/B978-0-323-90395-0.00013-9

Tamm—Horsfall Protein [14] or uromodulin (UMOD) [15,16] is the most abundant glycoprotein in human urine. This mature urinary protein is composed of 616 amino acids and has a deduced molecular weight of 67 kDa. Its amino acid composition displays 48 cysteines engaged in 24 disulfide bonds. The predicted structure of the mature protein includes four EGF-like domains Ca^{2+} binding; one cysteine-rich D8C domain; two zona pellucida domains (ZP domain); and one glycosylphosphatidylinositol-anchoring site at position 614. It has seven N-glycosylation sites representing an additional 30% of the final molecular weight (approximately 85—90 kD).

THP is synthesized by the epithelial cells of the thick ascending limb (TAL) of Henle's loop. After being transported to the apical side of the cell, it is released in the lumen of the nephron after proteolytic cleavage by hepsin [17]. As soon as it reaches the lumen, its ZP domains are free to interact, and the protein monomer associates and forms filaments of well-defined structure with an average length of 2.5 μm [18]. Filaments are organized in a three-dimensional network that can entrap influenza virus particles [19] as well as uEVs [1,10]. Independently from the purification protocol, there is always a co-segregation of uEVs and uromodulin, either as filaments or an unfolded protein.

Successful removal of THP is usually achieved with a combination of techniques such as ultracentrifugation, density gradient centrifugation, and size exclusion chromatography which are labor-intensive and time consuming for the operator.

2. Before you begin

- Sodium hydroxide 10 N stock solution (see Note 1)
  Weigh 4 g of sodium hydroxide and solubilize in a final volume of 10 mL with 0.1 μm filtered dH_2O (see Note 2).
- Silver chloride (0.1 mg/mL stock solution) (see Note 3)
  Weigh 2 mg of silver chloride and solubilize in a final volume of 20 mL with 0.1 μm filtered dH_2O.
- Sodium troclosene or sodium dichloroisocyanurate (4.5 mg/mL stock solution) (see Note 4)
  Weigh 9.0 mg of sodium troclosene and dissolve in a final volume of 20 mL with 0.1 μm filtered dH_2O.
- Uromodulin de-polymerization solution (see Note 5)
  Weigh urea (MW 60.06 g/mol) 54.05 g (9 M in 0.1 L).
    citric acid (MW 192.12 g/mol) 2.88 g (0.15 M in 0.1 L).
    L-arginine (MW 174.2 g/mol) 1.31 g (75 mM in 0.1 L).
    6-aminocaproic acid MW (131.17 g/mol) 0.98 g (75 mM in 0.1 L).
  Bring to 100 mL with 0.1 μm filtered dH_2O, stir to dissolve, check the pH, and if necessary adjust with a few drops of HCl to pH 3.2 (see Note 6).
- Citrate buffer stock solution
  Weigh citric acid (MW 192.12 g/mol) 19.12 g (1.0 M in 0.1 L).
  Weigh sodium citrate tribasic dihydrated (MW 294.10) 29.41 g (1.0 M in 0.1 L).

Bring each solution to 100 mL with 0.1 μm filtered dH$_2$O.
Titrate the sodium citrate solution with citric acid until you reach a pH of 7.4.
- Ethylenediaminetetraacetic acid disodium salt dihydrate stock solution
  Weight EDTA (MW 372.24 g/mol) 18.61 g (0.5 M in 0.1 L).
  Add 70 mL of 0.1 μm filtered dH$_2$O, stir to dissolve, and add dropwise 10 N
  NaOH to solubilize EDTA and reach the final pH of 7.4.
  Bring to the final volume of 100 mL with 0.1 μm filtered dH$_2$O.
- 4-(2-Hydroxyethyl)piperazine-1-ethanesulfonic acid, 10× HEPES buffer stock
  solution
  Weigh HEPES (MW 238.30 g/mol) 23.83 g (1.0 M in 0.1 L).
  Add 70 mL of 0.1 μm filtered dH$_2$O, stir to dissolve, and add dropwise NaOH
  10 N to solubilize HEPES and reach the final pH of 7.4.
  Bring to the final volume of 100 mL with 0.1 μm filtered dH$_2$O.
- HEPES washing solution
  10 mL of 1 M HEPES buffer.
  1 mL of silver chloride 0.1 mg/mL.
  1 mL of sodium troclosene 4.5 mg/mL.
  Bring to 1 L with 0.1 μm filtered dH$_2$O.
- 10× Ammonium bicarbonate.
  Weigh NH$_4$HCO$_3$ (MW 79.06 g/mol) 19.76 g (1.0 M in 0.15 L stock solution).
  Bring to 100 mL with 0.1 μm filtered dH$_2$O, and stir to dissolve.
- Trichloroacetic acid (TCA) precipitating solution (see Note 7)
  10 mL of TCA 100% or 6.1 N.
  40 mg of sodium deoxycholate (DOC).
  Add 40 mg of DOC to 10 mL of TCA in a 15 mL tube. Seal and carefully mix until
  DOC is dissolved. Store at 4°C.

Notes

1. Sodium hydroxide needs to be handled with extreme caution as it is very corrosive and it may cause severe burns and/or serious permanent eye damage. It is extremely harmful if accidentally ingested, or through skin contact and/or by dust inhalation.

2. Solubilization of NaOH is exothermic and must be performed with extreme care since it is highly caustic.

3. Silver chloride solubility is very low = 2 mg/mL in water. Prepare the solution the day before and leave it to resolubilize overnight (ON) in an end-over-end tube rotator or magnetic stirrer.

4. Sodium dichloroisocyanurate in powder form is toxic to inhale, on the skin and harmful if swallowed. Handle it with protective gloves under a fume hood or wear a face shield or eye protection in combination with breathing protection.

5. Prepare the solution the day before. 9 M urea is very close to the limit of solubility. Stir overnight. Solubilization of urea is endothermic. Do not warm up to avoid dissociation of urea into ammonia and isocyanic acid (protein carbamylation).

6. Hydrochloric acid is corrosive and concentrated hydrochloric acid should be handled under a fume hood while wearing safety glasses, gloves, and a laboratory coat.

7. TCA is a very strong acid, both toxic and caustic. It may cause skin and eye burns. Wear gloves and safety glasses while handling it.

3. Key resources table

Reagent or resource	Source	Identifier
Antibodies		
Rabbit anti-tumor susceptibility gene 101 (TSG101)	Sigma	T5826
Goat anti-angiotensin converting enzyme 2 (ACE2)	R&D System	AF933
Rabbit anti-Syntenin-1 (EPR8102)	Abcam	ab133267
Biotynilated mouse anti CD9 (HI9a)	Biolegend	312112
IRDye 800CW Donkey anti-Rabbit IgG	Li-Cor Biosciences	926-32213
IRDye 800CW Donkey anti-Goat IgG	Li-Cor Biosciences	926-68074
IRDye 800CW Streptavidin	Li-Cor Biosciences	926-32230
Biological samples		
Urine	Human	Healthy
Chemicals, peptides, and recombinant proteins		
6-aminocaproic acid	Sigma—Aldrich	A7824-100G
Acetone	Sigma—Aldrich	270725-1L
Ammonium bicarbonate	Sigma—Aldrich	09830-500G
Chloroform	Sigma—Aldrich	288306-100ML
Citric acid	Sigma—Aldrich	251275-500G
EDTA	Bio-Rad	1610729
L-arginine	Sigma—Aldrich	A5006-100G
HEPES	Sigma—Aldrich	H23830-100G
Intercept (PBS) blocking buffer	Li-Cor Biosciences	927-70001
Methanol	Sigma—Aldrich	179337-1L
Silver chloride	Sigma—Aldrich	227927-10G
Sodium deoxycholate	Sigma—Aldrich	30970-25G
Sodium dichloroisocyanurate	Sigma—Aldrich	35915-50G
Sodium citrate tribasic dihydrated	Sigma—Aldrich	S4641-500G
Sodium hydroxide	Sigma—Aldrich	S5881-500G
Trichloroacetic acid	Sigma—Aldrich	T0699-100ML
Urea	Bio-Rad	1610731

4. Materials and equipment

- Benchtop centrifuge
- Separating funnel
- Funnel
- Dialysis membrane MWCO 1000 kDa width 16 mm, 0.79 mL/cm (Repligen code # 131486) and/or width 31 mm 3.1 mL/cm (Repligen code # 131492).
- Universal fasteners for all membrane types 50 mm width (Repligen code # 131486)
- Dialysis membrane MWCO 3.5 kDa (Repligen code # 1327200
- 0.1 μm syringe filters (Sartorious 16553K)
- Parafilm
- Vacuum concentrator
- End-over-end rotor or a rocker

5. Step-by-step method details

1. Centrifuge urine with a bench top centrifuge at $2000 \times g$ for 30 min at room temperature (see Note 8).
2. Pour urine SN $2000 \times g$ into a graduated cylinder and record the volume.
3. Add 5.0 mL of citrate buffer (50 mM final concentration) and 1.0 mL of EDTA solution (5 mM final concentration) for every 93.8 mL of urine.
4. Add 0.1 mL of silver chloride (0.1 mg/L final concentration) and 0.1 mL of troclose solution (0.1 mg/L final concentration) for every 93.8 mL of urine.
5. Set up the Hydrostatic Filtration Dialysis system as follows:
   a. Cut a piece of the dialysis tube approximately the same length as from the bottom of the cylinder to the top of the separating funnel (Fig. 7.1A1 (16 mm) and A2 (31 mm).
   b. Insert one end of the dialysis tube into the top of a separating funnel (Fig. 7.1B1 for the 16 mm membrane) or a funnel (Fig. 7.1C1 for the 31 mm membrane).
   c. Seal and tighten the dialysis membrane around the pipe of the funnel with a strip of parafilm (Fig. 7.1B2 and C2).
   d. Seal the bottom end of the dialysis tube with a universal dialysis clip (Fig. 7.1D1 and D2) (see Note 9).
   e. Pour 200 mL of filtered 0.1 μm dH$_2$O into the funnel to make sure the system is fully sealed.
   f. Get rid of any entrapped air in the dialysis tube by gently pressing on the dialysis membrane (Fig. 7.1E3 arrows).
   g. Close the valve of the separating funnel and apply some pressure on the dialysis tube by gently squeezing the membrane between fingers. Check if the dialysis clip is well tightened and that the 0.1 μm filtered dH$_2$O does

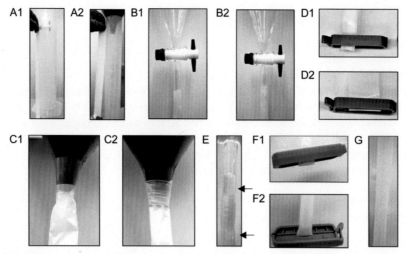

FIGURE 7.1

Hydrostatic filtration dialysis; setting up the system. A1—A2: the length of the dialysis membrane is measured and cut to fit in the cylinder. B1—C1: inserting the membrane dialysis tubes in the pipe of a separating funnel/regular funnel. B2—C2; tighten the membrane with a strip of parafilm. D1—D2; opposite ends are sealed with a universal dialysis tube closure. After pouring 200 mL of dH$_2$O into the funnel, inspect for the presence of bubbles of air trapped in the tube (E arrows). Sealing of the bottom end: water should not leak through the bottom of the dialysis tube (F1) but be collected on top (F2). G; applied pressure pushes water up through the membrane.

not flowthrough the end of the dialysis membrane (Fig. 7.1F). Please note that when you push the dialysis membrane some water will run through the dialysis membrane (Fig. 7.1G) and collect on top of the clip (Fig. 7.1F2). If you use the larger dialysis membrane (31 mm), apply some pressure with your fingers and observe if there is some water spurting at both ends, at the point of insertion of the dialysis tube in the funnel and at the bottom where the clip seals the tube of the dialysis membrane.

   h. Open the clip, let the water flow into the sink, and seal it back again.

6. Pour the urine into the funnel (Fig. 7.2A1 and A2) and check if any air is entrapped in the dialysis tube (Fig. 7.2A1 magnification, arrows), particularly when using the narrow size membrane (16 mm).

   a. The membrane will start to leak urine (Fig. 7.2B1 and B2) which will pool at the bottom of the cylinder (Fig. 7.2C1 and C2).

   b. Let the urine concentrate up to a few mL (Fig. 7.2D1 and D2).

   c. Refill the separating funnel with 200 mL of HEPES washing buffer and let it reconcentrate up to a few mL (Fig. 7.2E1 and E2). Note that the acronym

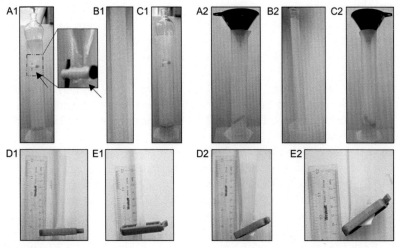

FIGURE 7.2

Hydrostatic filtration dialysis. A1—A2; the separating funnels are filled in with urine supernatant after centrifugation at 2000 g. Entrapped air bubbles in the tube are removed by gentle pressure on the dialysis tube (A1 magnification). B1—B2; the pressure of urine in the funnel and the column of the dialysis tube pushes urine through the dialysis membrane meshwork. C1—C2; urine—along with molecules smaller than the MWCO (HDb)—drip into the cylinder, pooling at the bottom. D1—D2; when the urinary solution reaches 7—10 cm in height, the funnel is refilled with HEPES washing buffer which flushes away the leftovers. E1—E2; the uEV solution is concentrated up to the desired final volume.

for the retentate is HFDa (where "a" stands for "above" the MWCO) and HFDb (where "b" stands for "below" the MWCO) for the flowthrough.

7. Remove the uEV fraction (HFDa) opening the clip with the end of the dialysis tube on top of a 50 mL tube for an easy recovery to avoid spillover of any precious material. Carefully squeeze down all the solution into the 50 mL tube with your fingers. You might see some gelatinous yellowish solution being "milked" into the 50 mL tube. This is made of THP polymers which are adsorbed in the internal surface of the dialysis tube (protein fouling). Samples can be stored or processed with urea to remove THP. If you proceed with the denaturation step, you can use the same HFD system.
   Notes

8. Depending on the geometry of the rotor and its braking design, the very first pellet made of cast of THP polymers might detach from the bottom of the tube and float in the urine. For this reason, very low or no braking settings are highly recommended.

9. At the bottom end, the last cm of the dialysis membrane could be folded over to offer some more thickness for a better sealing pressure of the clip (Fig. 7.1D2).

8. Thaw the HFDa sample or continue from step 7. Work out the final volume and dilute proportionally with two volumes of urea solution (urea final concentration 6 M).

   a. Place the tube in an end-over-end rotor (or a rocker) and incubate for 30 min.

   b. Reload the separating funnel and top it up with 50 mL of 6 M urea solution to wash away any depolymerized uromodulin left in the retentate solution.

   c. Wash out the excess of urea and THP from the retentate with $1\times$ ammonium bicarbonate solution.

9. Work out the protein concentration.

10. Concentrate by vacuum concentrator or precipitate the protein by TCA-DOC [20] for the equivalent volume of 20 μg of total protein.

    a. Add 250 μL of TCA/DOC solution for each mL of protein solution. TCA final concentration has to be 20%.

    b. Vortex and incubate on ice for 30 min.

    c. Centrifuge at max speed in a refrigerated microcentrifuge for 30 min at 4°C. Discard the supernatant.

    d. Add 0.8–1.0 mL of 100% (v/v) acetone (see Note 10), and incubate at −20°C from 2 h to overnight.

    e. Centrifuge at max speed in a refrigerated microcentrifuge for 30 min at 4°C. Discard the supernatant.

    f. Let the pellet dry in the fume hood for 10 min.

11. Extract lipid and precipitate protein by the chloroform–methanol protocol [21].

    a. Add 400 μL of 100% (v/v) methanol (see Note 11) to 100 μL of the sample. Vortex thoroughly and centrifuge for 30 s at $9000 \times g$.

    b. Add 200 μL of chloroform (see Note 12). Vortex vigorously and centrifuge for 30 s at $9000 \times g$.

    c. Add 300 μL of deionized water, mix vigorously, and centrifuge for 5 min at $9000 \times g$.

    e. Discard the aqueous upper phase and leave the interface protein layer untouched.

    f. Add 300 μL of 100% (v/v) methanol and centrifuge for 10 min at max speed.

    g. Allow it to dry for 10 min under the fume hood.

12. SDS-PAGE and Western BlotWestern Blot for quality controls were performed according to Musante et al. [1].

    Notes

10. Acetone is toxic and highly flammable. Avoid flames and handle it under a fume hood.

11. Methanol is toxic and highly flammable. Avoid flames and handle it under a fume hood.

12. Chloroform is harmful and should be handled under a fume hood.

6. Expected outcomes

The protein composition of the fractions can be analyzed by gel electrophoresis and Western blot using antibodies against some of the protein markers characteristic of the EV proteome reflecting EV topology, such as transmembrane proteins— tetraspanin CD9 antigen (CD9_Human UniProtKB-P21926) and Angiotensin-converting enzyme 2 (ACE2_Human UniProtKB-Q9BYF1)—or intraluminal EV markers—Tumor susceptibility gene 101 protein (TSG101_Human UniProtKB-Q99816) or Syntenin-1 (SDCBP1-Human UniProtKB-O00560). Colloidal Coomassie staining of the fractions obtained after concentration by HFD and denaturation of THP by urea, followed by a second HFD step are shown in Fig. 7.3A and B, representing two independent uEV preparations from different specimens. HFD-enriched uEVs (HFDa) as well as relatively large amount of THP (*) (Fig. 7.3A and B lanes 1,2 and a) whose final yield is different from urine to urine are shown in two separate urine collections in Fig. 7.3A and B. After unfolding the THP polymer by urea, all the monomers and/or oligomers smaller than the MWCO of 1000 kDa can be filtered and recovered in the flowthrough (HFDb-urea). Fig. 7.3A and B: lanes 4,5 and b. While uEVs are retained in the tube (HFDa-urea) Fig. 7.3A and B: lanes 3 and c; Western blot detection of EVs markers showed a consistent recovery of the signal in the retentate particularly for transmembrane proteins such as CD9 (Fig. 7.3C and ACE2 Fig. 7.3E). On the other hand, some signal is detectable in the flowthrough for TSG101 (Fig. 7.3D) and syntenin-1 (SDCBP) (Fig. 7.3D). Quality control of the samples ready for Mass Spectrometry analysis (Fig. 7.3B) shows that after delipidation by chloroform-methanol, the total amount of protein is recovered in the pellet (Fig. 7.3B lane d) with a protein pattern equal to the fraction without any lipid extraction (Fig. 7.3B lane c).

7. Advantages

Hydrostatic filtration dialysis (HFD) is a simple system that requires minimal financial investment [22]. It can safely process large volumes of urine without the need for any sophisticated instruments requiring highly specialized training. HFD captures the majority of uEVs with minimal substantial loss and/or fractionation—which is a problem affecting the differential centrifugation, density gradient, and size exclusion chromatography protocols. Thus, the key feature in implementing HFD resides in sample preparation, which needs to be very simple and follow a very limited number of steps to achieve an optimal EV prep where the majority of the uEV population is represented. Multiple samples can be processed at the same time and in such a number that can match or even be higher than the ultracentrifugation rotor tube cavities, generally from six to eight places, depending on the type and nominal max volume capacity of the rotor in use. The average filtration rate is 50—80 mL per hour for the 16 mm membrane and 150—200 mL per hour for the 31 mm membrane.

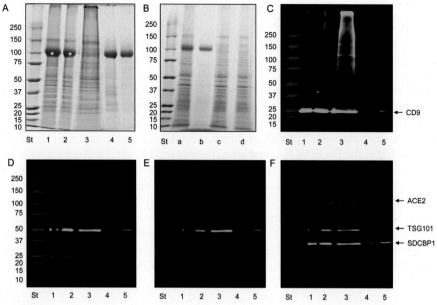

FIGURE 7.3

SDS-polyacrylamide gel electrophoresis (SDS-PAGE) and WB analysis of uEVs enriched by HFD before and after depletion of THP. Panel A—Colloidal Coomassie staining of lane 1 HFDa (loaded with 20 µg); lane 2, HFDa (loaded with 15 µg); lane 3, HFDa urea (loaded with 15 µg); lane 4, HFDb urea from the first 10 mL of the flowthrough (loaded with 15 µg); lane 5, HFDb urea from the following 40 mL of the flowthrough (loaded with 15 µg). Panel B—Colloidal Coomassie staining of the fractions obtained from a second sample of urine enriched by HFD. Lane a, HFDa (loaded with 15 µg); lane b, HFDb urea from the first 10 mL of the flowthrough (loaded with 15 µg); lane c, HFDa urea (loaded with 15 µg); lane c, HFDa urea (loaded with 15 µg) after delipidation with chloroform methanol. Panel C: WB detection of CD9. Gel loaded as per panel A. Samples were solubilized in an electrophoresis buffer without DTT (anti CD9 1 µg/mL). Panel D—WB detection of TSG101 (0.5 µg/mL). Gel loaded as per panel A. Samples were solubilized in an electrophoresis buffer with 50 mm DTT. Panel E—Nitrocellulose membrane in D was re-probed with anti-ACE2 (1 µg/mL). Panel F—Nitrocellulose membrane in D and E was reprobed with anti-syntenin-1 (0.5 µg/mL). Primary antibodies were diluted with an Odyssey blocking solution in PBS and 0.15% (v/v) Tween-20 and incubated overnight at RT. Secondary antibodies were conjugated with red (displayed in red excitation 680 nm, emission 700 nm) or infrared (displayed in green color excitation 780 nm, emission 800 nm). Dye-coupled were diluted in an Odyssey blocking buffer in PBS 0.02 µg/mL and incubated for 1 h at RT. Acquisition of the fluorescent signal was performed by Odyssey Infrared Imaging system with a resolution set at 169 µm (LI-COR Biosciences). Molecular weights are expressed in kilo Daltons (kDa). Protein loading based on Bradford Coomassie protein assay.

Furthermore, the same HFD system can be used when denaturation of the THP polymers follows the enrichment step, thus saving time and resources (same dialysis membrane).

8. Limitations

HFD performs at its best with large volumes of urine (>50 mL). Although it can be useful also when handling smaller volumes, it does not offer the same advantages in terms of high volume/sample processing time ratio. Moreover, with small volumes, the uEVs nonspecific adsorption on the inner surface of the dialysis tube can result in an important loss of material, particularly after removing THP. Decreasing the length of the membrane to achieve better recovery of uEVs decreases the overall filtration surface, therefore proportionally increasing the sample processing time. The filtration rate is strongly dependent on the volume of urine. Hence, it is fast at the very beginning and it will slow down proportionally in parallel with the decrease of the filtered volume.

9. Optimization and troubleshooting
9.1 Dialysis membrane and urine filtration time

HFD is an easy system to process large volumes of urine to enrich EVs of any size. However, some considerations are necessary on the nature of the system and the composition of its parts. Firstly, the dialysis membrane is relatively fragile, it can easily be pierced by the tip of the pipe of the separating funnel. Moreover, some little cuts might occur while sealing the end of the dialysis tube with the clip. As there are several types of clips that can be used, it is important to choose those which do not pin too tight as they can sever the membrane. Holes and cuts on the dialysis membrane can be very small and difficult to see but they would nonetheless allow a preferential way out to the solution leading to loss of material. Secondly, the filtration rate is strictly dependent on the volume of the solution contained in the separating funnel, so it can be relatively fast at the beginning and then it can progressively slow down proportionally to a lower volume of urine. Moreover, there are variations in the filtration rate from roll to roll and sometimes also within the same roll, as different portions of the membrane can perform differently. Sometimes differences in the membrane are even visible to the naked eye, as some segments of the dialysis membrane can vary in thickness and/or transparency. This might depend on the manufacturing process. Lastly, as happens with any other filtration system, protein fouling can occur, thus contributing to a poor filtration rate.

The main protein adsorbed on the inner surface of the dialysis membrane is Tamm—Horsfall Protein (THP) in its polymeric forms [23]. Therefore, the amount of leftover THP after the initial centrifugation step at $2000 \times g$, together with the ionic composition and ionic strength of the urine are all contributing factors that modulate the rate of THP polymerization and contribute to a variation in the filtration rate from sample to sample, as you can see in Fig. 7.3 panel A and B representing the final yield of THP recovered in the retentate (HFDa) for two different specimens.

9.2 Urea denaturation and uEVs integrity

THP depolymerization is achieved with 6 M urea concentration in acid conditions, pH < 3.2 [12]. It cannot be ruled out that such harsh conditions could destabilize the integrity of the phospholipidic membrane bilayers in part of the uEV heterogeneous populations. This can lead to ruptures in the phospholipidic membrane with the consequent release of some of their internal cargo outside of the vesicle. Western blot detection of the two intraluminal EV markers TSG101 and syntenin-1 (Fig. 7.3D and F) shows that some signal is present in the flowthrough of the 50 mL of urea solution used to flush away any unfolded THP. On the other hand, some membrane protein markers—such as tetraspanin CD9 with its four transmembrane domains and ACE2 with one transmembrane domain—are retained in the HFDa-urea fraction with no signal detected in the flowthrough HFDb-urea (Fig. 7.3C and E). Conversely, we cannot exclude that the presence of urea can increase the polydispersity and favor disaggregation of uEV clumps with some small EVs crossing the dialysis membrane meshwork. The use of a dialysis membrane with a smaller MWCO (300 kDa) might reduce the loss of EVs. However, we have to keep in mind that the filtration gravity is dramatically reduced by the use of a narrower MWCO membrane and therefore some additional gravity forces such as using, e.g., a Vacuum Dialysis system [24] would need to be applied (Fig. 7.4). Fine tuning of the vacuum force is pivotal to prevent the dialysis membrane from bursting.

9.3 Potential solution to optimize the procedure

To minimize differences in the ionic composition of distinct urinary specimens, samples can be diluted with dH_2O, which favors THP depolymerization [25]. However, this increases the sample volume and extends the time to enrich uEV proportionally. Alternatively, 1 mM $ZnSO_4$ favors THP polymerization, and the extension of filaments can increase the recovery of THP in the first centrifugation step at 2000 × g [26]. Furthermore, supplementation with $ZnSO_4$ seems to have a protective role in sample long-term storage at both −20 and −80°C. A sodium chloride precipitation according to Tamm and Horsfall (1950) could be taken into consideration but it involves overnight sample incubation at 4°C. Finally, adsorption of THP onto a cake of diatomaceous earth filter could be performed to remove the excess of THP [27]. However, this step can extend the processing time, while implementation with the Büchner filter and vacuum flask can limit the number of samples processed simultaneously.

FIGURE 7.4

Vacuum dialysis. The separating funnel is inserted into a bored stopper of the same diameter as the pipe of the separating funnel. The dialysis membrane with a 300 kDa MWCO is assembled and sealed as per HFD instructions. Gentle vacuuming is applied to avoid ruptures of the dialysis membrane.

10. Safety considerations and standards

Hydrostatic filtration dialysis is a safe technique that does not require neither the use of any hazardous chemicals nor specialized equipment or high expertise. The steps to set up the system are easy to follow and the operator does not have to perform any hazardous steps but pour urine into the funnel making sure to avoid any spillovers. Finally, the chemicals used to unfold THP do not represent any hazard, as shown in the pictograms on the chemical containers. The only chemical substance which might require some attention during the preparation of the stock solution— particularly when handled in bulk—is sodium troclosene or sodium dichloroisocya- nurate, which is normally used as a bactericide. Alternatively, purifying water tablets can be used, which are available for purchase at any drugstore or top online retailers.

11. Alternative methods/procedures

To concentrate uEVs as a first step, differential centrifugation, salting out precipitation with polyethylene glycol (PEG), and ultrafiltration using centrifugation devices have been proposed. The presence of THP in the concentrated fraction still requires the addi- tion of a denaturation step such as DTT reduction and a second step to separate uEV from the monomeric unfolded THP. In this regard, the best choice would be size- exclusion chromatography, preferred to ultracentrifugation because unfolded THP keeps on precipitating as reported in the original publication [10]. Density gradient ul- tracentrifugation is another method to separate THP and uEVs. However, it adds some additional labor-intensive work such as long centrifugation times, dilution of the den- sity fraction, and reconcentration by ultracentrifugation to recover uEV in manageable volumes. A less common system to process relatively large volumes and concentrate EVs depleted from soluble contaminants is Tangential Flow Filtration (TFF) but the same considerations for THP co-isolation need to be taken into account.

References

[1] Musante L, Bontha SV, La Salvia S, Fernandez-Piñeros A, Lannigan J, et al. Rigorous characterization of urinary extracellular vesicles (uEVs) in the low centrifugation pellet — a neglected source for uEVs. Sci Rep 2020;10(1):3701. https://doi.org/ 10.1038/s41598-020-60619-w.

[2] Erdbrügger U, Blijdorp CJ, Bijnsdorp IV, Borràs FE, Burger D, et al. Urinary extracel- lular vesicles: a position paper by the Urine Task Force of the International Society for Extracellular Vesicles. J Extracell Vesicles 2021;10(7):212093. https://doi.org/10.1002/ jev2.12093.

[3] Pisitkun T, Shen RF, Knepper MA. Identification and proteomic profiling of exosomes in human urine. Proc Natl Acad Sci U S A 2004;101(36):13368—73. https://doi.org/ 10.1073/pnas.0403453101.

[4] Gonzales PA, Pisitkun T, Hoffert JD, Tchapyjnikov D, Star RA, et al. Large-scale proteomics and phosphoproteomics of urinary exosomes. J Am Soc Nephrol 2009;20(2): 363–79. https://doi.org/10.1681/ASN.2008040406.

[5] Wu Q, Poulsen SB, Murali SK, Grimm PR, Su XT, et al. Large-scale proteomic assessment of urinary extracellular vesicles highlights their reliability in reflecting protein changes in the kidney. J Am Soc Nephrol 2021;32(9):2195–209. https://doi.org/10.1681/ASN.2020071035.

[6] Eldh M, Mints M, Hiltbrunner S, Ladjevardi S, Alamdari F, et al. Proteomic profiling of tissue exosomes indicates continuous release of malignant exosomes in urinary bladder cancer patients, even with pathologically undetectable tumour. Cancers 2021;13:3242.

[7] Dhondt B, Geeurickx E, Tulkens J, Van Deun J, Vergauwen G, et al. Unravelling the proteomic landscape of extracellular vesicles in prostate cancer by density-based fractionation of urine. J Extracell Vesicles 2020;9(1):1736935. https://doi.org/10.1080/20013078.2020.1736935.

[8] Erozenci LA, Pham TV, Piersma SR, Dits NFJ, Jenster GW, et al. Simple urine storage protocol for extracellular vesicle proteomics compatible with at-home self-sampling. Sci Rep 2021;11(1):20760. https://doi.org/10.1038/s41598-021-00289-4.

[9] Mitchell PJ, Welton J, Staffurth J, Court J, Mason MD, Tabi Z, Clayton A. Can urinary exosomes act as treatment response markers in prostate cancer? J Transl Med 2009;7: 1–3.

[10] Fernández-Llama P, Khositseth S, Gonzales PA, Star RA, Pisitkun T, Knepper MA. Tamm-Horsfall protein and urinary exosome isolation. Kidney Int 2012;77(8): 736–42. https://doi.org/10.1038/ki.2009.550.

[11] Liu X, Chinello C, Musante L, Cazzaniga M, Tataruch D, et al. Intraluminal proteome and peptidome of human urinary extracellular vesicles. Proteomics Clin Appl 2015; 9(5–6):568–73. https://doi.org/10.1002/prca.201400085.

[12] Xu X, Barreiro K, Musante L, Kretz O, Lin H, et al. Management of Tamm-Horsfall protein for reliable urinary analytics. Proteomics Clin Appl 2019;13(6):e1900018. https://doi.org/10.1002/prca.201900018.

[13] Maggio S, Polidori E, Ceccaroli P, Cioccolon A, Stocchi V, Guescini M. Current methods for the isolation of urinary extracellular vesicles. Methods Mol Biol 2021; 2292:153–72. https://doi.org/10.1007/978-1-0716-1354-2_14.

[14] Tamm I, Horsfall Jr FL. Characterization and separation of an inhibitor of viral hemagglutination present in urine. Proc Soc Exp Biol Med 1950;74(1):106–8.

[15] Muchmore AV, Decker JM. Uromodulin: a unique 85-kilodalton immunosuppressive glycoprotein isolated from urine of pregnant women. Science 1985;229(4712): 479–81. https://doi.org/10.1126/science.2409603.

[16] Pennica D, Kohr WJ, Kuang WJ, Glaister D, Aggarwal BB, Chen EY, Goeddel DV. Identification of human uromodulin as the Tamm-Horsfall urinary glycoprotein. Science 1987;236(4797):83–8. https://doi.org/10.1126/science.3453112.

[17] Devuyst O, Olinger E, Rampoldi L. Uromodulin: from physiology to rare and complex kidney disorders. Nat Rev Nephrol 2017;13(9):525–44. https://doi.org/10.1038/nrneph.2017.101.

[18] Stanisich JJ, Zyla DS, Afanasyev P, Xu J, Kipp A, et al. The cryo-EM structure of the human uromodulin filament core reveals a unique assembly mechanism. Elife 2020; 20(9):e60265. https://doi.org/10.7554/eLife.60265.

[19] Bayer ME. An electron microscope examination of urinary mucoprotein and its inter-action with influenza virus. J Cell Biol 1964;21(2):265−74. https://doi.org/10.1083/jcb.21.2.265.

[20] Koontz L. TCA precipitation. Methods Enzymol 2014;541:3−10. https://doi.org/10.1016/B978-0-12-420119-4.00001-X.

[21] Wessel D, Flügge UI. A method for the quantitative recovery of protein in dilute solu-tion in the presence of detergents and lipids. Anal Biochem 1984;138(1):141−3. https://doi.org/10.1016/0003-2697(84)90782-6.

[22] Musante L, Tataruch D, Gu D, Benito-Martin A, Calzaferri G, Aherne S, Holthofer H. A simplified method to recover urinary vesicles for clinical applications, and sample banking. Sci Rep 2014;4. https://doi.org/10.1038/srep07532.

[23] Zumstein L. Dialysis and ultrafiltration. Curr Protoc Mol Biol 2001. https://doi.org/10.1002/0471142727.mba03cs41. Appendix 3: Appendix 3C.

[24] Puhka M, Nordberg ME, Valkonen S, Rannikko A, Kallioniemi O, Siljander P, Af Hällström TM. KeepEX, a simple dilution protocol for improving extracellular vesicle yields from urine. Eur J Pharm Sci 2017;15(98):30−9. https://doi.org/10.1016/j.ejps.2016.10.021.

[25] Liu Z, Cauvi DM, Bernardino EMA, Lara B, Lizardo RE, et al. Isolation and character-ization of human urine extracellular vesicles. Cell Stress Chaperones 2018;23(5):943−53. https://doi.org/10.1007/s12192-018-0902-5.

[26] Cavallone D, Malagolini N, Frascà G, Stefoni S, Serafini-Cessi F. Salt-precipitation method does not isolate to homogeneity Tamm-Horsfall glycoprotein from urine of pro-teinuric patients and pregnant women. Clin Biochem 2002;35(5):405−10. https://doi.org/10.1016/s0009-9120(02)00329-6.

[27] Tataruch-Weinert D, Musante L, Kretz O, Holthofer H. Urinary extracellular vesicles for RNA extraction: optimization of a protocol devoid of prokaryote contamination. J Extracell Vesicles 2016;24(5):30281. https://doi.org/10.3402/jev.v5.30281.

Sample preparation for proteomics and mass spectrometry from animal samples

8

Alessio Di Luca and Giuseppe Martino

Department of Bioscience and Agro-Food and Environmental Technology, University of Teramo, Teramo, Italy

1. Introduction

Proteomics consists of the high-throughput study of the proteome (the protein complement of a genome). The proteome, unlike the genome, varies with time and is subject to several changes of a different nature, which are determined by the interactions of the different molecules and by environmental factors [1,2]. Due to these features, proteomics is ideally suited to enhancing our knowledge on animal biology. Indeed, proteomic analysis that investigates large numbers of proteins for their associations with physiological and pathological conditions may contribute to a better understanding of the biological processes underpinning them. Moreover, it is suited to the search for and evaluation of new marker proteins that can be used, in animal biology, as early predictors of animal welfare, meat quality, etc. [3].

For many years, SDS PAGE and two-dimensional gel electrophoresis (2D PAGE) were the main separative techniques used to relate protein abundance pattern to the animal biology queries [4,5]. In more recent times, the development of mass spectrometric analyses and sophisticated software allowed improved separation and identification of the proteins of interest [6]. Among them, label-free LC-MS (liquid chromatography mass spectrometry), a form of "shotgun proteomics," is a method for measuring peptide concentrations in complex samples using a combination of high-performance liquid chromatography (HPLC) and MS. This technique is capable of analyzing different specimens, is cost-effective, the samples are straightforward to process, is reproducible, and can be scaled to a large number of samples. As a result, the technique can be applied for the investigation of a wide range of tissue and biofluids specimens that can be used in animal proteomic experiments [7,8].

Proteins extraction from biological material and the subsequent conversion to peptides suitable for MS-based proteome analysis are pivotal for proteomic studies. Here, we present our routine method for protein extraction, an exudate collected postmortem from muscle tissue following centrifugation. This substrate is straightforward to prepare, reproducible, rich in proteins and its composition is not

Proteomics Mass Spectrometry Methods. https://doi.org/10.1016/B978-0-323-90395-0.00009-7

predominated by a small number of highly abundant proteins, which make it possible to identify also less abundantly expressed proteins [4]. We also present a comprehensive method for protein purification and digestion, using Filter-Aided Sample Preparation (FASP) for animal specimens [9,10]. This method uses standard ultrafiltration devices for an efficient removal of impurities and detergent to enable subsequent proteome analysis with a more complete coverage of the proteome [11].

2. Materials and equipment

Prepare all solutions using LC-MS grade water and all chemicals must be of the highest purity. Prepare and store all reagents at room temperature (unless otherwise indicated). Diligently follow all waste disposal regulations when disposing waste materials. Minimize handling and manipulations of samples.

2.1 Equipment

1. Ultracentrifuge capable of spinning centrifuge tubes (25 × 89 mm) at 3400/ 6000 ×g at 4 °C.
2. Spinning centrifuge tubes (25 × 89 mm) or 50 mL polypropylene centrifuge tubes.
3. Microcentrifuge tubes (different volumes).
4. Cuvette or Microplate Spectrophotometer reading 96-well plate format at 595 nm.
5. Absorbance reader.
6. Microcon-10 kDa Centrifugal Filter Unit with Ultracel-10 membrane (Amicon Ultra-0.5 Centrifugal Filter Devices; Merck Millipore) or similar device.
7. Microcentrifuge capable of spinning 1.7 mL microcentrifuge tubes at 14,000 ×g at 20°C.
8. Vacuum Evaporator.
9. Pierce C18 Spin Columns (e.g., Thermo Fisher Scientific).
10. Dry Heating Block.
11. pH meter.
12. Vortex.
13. Fume Hood.
14. Ultimate 3000 RSLCnanoLC system (Thermo Fisher Scientific).
15. Orbitrap Fusion Tribrid mass spectrometer (Thermo Scientific).
16. SilicaTip Standard Coating Tubing OD/ID 360/20 μm Tip, ID 10 μm, length 5 cm (New Objective).
17. Trapping column PepMap100, C18, 300 μm × 5 mm, 5 μm particle size, 100 Å pore size (Thermo Scientific).
18. Analytical column Acclaim PepMap 100, 75 μm × 50 cm, 3 μm bead diameter column (Thermo Scientific).

19. Proteome Discoverer v.2.2 (Thermo Fisher Scientific) with the Sequest HT algorithm and Percolator, suitable protein database for the animal under study (e.g., Sus Scrofa).

2.2 Reagents

1. Bradford assay kit.
2. Bovine serum albumin.
3. LC-MS grade water.
4. Urea.
5. Thiourea.
6. Trizma HCl.
7. Trizma Base.
8. CHAPS Detergent (3-((3-cholamidopropyl) dimethylammonio)-1-propanesulfonate).
9. Sodium dodecyl sulfate (SDS).
10. DL-Dithiothreitol (DTT).
11. Iodoacetamide (IAA).
12. Ammonium Bicarbonate.
13. Sodium Chloride (NaCl).
14. Sodium hydroxide (NaOH).
15. Sequencing-grade modified trypsin (e.g., Promega, Sigma Aldrich).
16. ProteaseMax Surfactant Trypsin Enhancer (Promega).
17. Trifluoroacetic acid (TFA).
18. Acetonitrile (ACN).
19. Formic acid.

3. Before you begin

3.1 Buffer preparation for protein quantification and FASP

1. Lysis Buffer: 7 M Urea, 2 M Thiourea, 30 mM Tris, 4% CHAPS—pH 8.5. To make a 50 mL solution of the lysis buffer add 21.021 g of Urea, 7.612 g of Thiourea, 0.18171 g Trizma HCl and 2 g of CHAPS to 50 mL LC-MS Water. Mix and adjust pH to 8.5 with HCl if required (see Note 1). Dispense into microfuge tubes in aliquots (e.g., 1 mL, 2 mL), label clearly and store at −20°C (see Note 2).
2. 0.5 M DL-Dithiothreitol (DTT): 0.0077 g of DTT made up in 100 μL of freshly prepared 50 mM ammonium bicarbonate.
3. 8 M Urea solution: To make a 5 mL solution of the 8 M Urea solution add 2.4 g Urea to 5 mL of LC-MS grade water (see Note 3).
4. 0.1 M Tris-HCl stock solutions (pH 8.5 and pH 7.9): these solutions are used to prepare the wash buffers for FASP (50 mL). To make a 50 mL solution of the Tris-HCl pH 8.5 add 0.221 g Trizma HCl and 0.436 g Trizma Base to 50 mL

LC-MS Water. To make a 50 mL solution of the Tris-HCl pH 7.9 add 0.488 g Trizma HCl and 0.230 g Trizma Base to 50 mL LC-MS Water. Make aliquots and store at 4°C up to 6 months (see Note 4).

5. Wash buffer 1: A solution of 8 M urea in 0.1 M Tris—HCl pH 8.5. To make a 5 mL solution of the Wash buffer 1 add 2.4 g Urea to 5 mL of 0.1M Tris-HCl stock solutions pH 8.5 (see Note 3).

6. Wash buffer 2: A solution of 8 M urea in 0.1 M Tris—HCl pH 7.9. To make a 5 mL solution of the Wash buffer 1 add 2.4 g Urea to 5 mL of 0.1M Tris-HCl stock solutions pH 7.9 (see Note 3).

7. Alkylation solution: A solution of 0.05 M Iodoacetamide in Wash buffer 1 (8 M Urea in 0.1 M Tris-HCl pH 8.5). To prepare 5 mL of this solution add 0.046 g of Iodoacetamide to 5 mL of Wash buffer 1 (8 M Urea in 0.1 M Tris-HCl pH 8.5) (see Note 3).

8. Rinse solution: to make a 5 mL solution of the 0.5 M Rinse solution add 0.146 g NaCl to 5 mL of LC-MS grade water (see Note 3).

3.2 Trypsin digestion and peptide extraction

1. 50 mM Ammonium Bicarbonate: to make a 5 mL solution of the 50 mM Ammonium Bicarbonate, add 0.0198 g Ammonium Bicarbonate to 5 mL of LC-MS grade water (see Note 3).

2. Trypsin reconstitution buffer: 50 mM acetic acid.

3. Trypsin reconstitution: Reconstitute the lyophilized trypsin in trypsin reconstitution buffer to give a final concentration of 1 μg/μL. Dispense into microfuge tubes in aliquots, label clearly and store at −20°C until required (see Notes 5 and 6).

4. ProteaseMax Surfactant, Trypsin Enhancer: Add 100 μL of freshly prepared 50 mM ammonium bicarbonate (pH~7.8) to a vial of ProteaseMAX Surfactant to give a 1% solution. Dispense into microfuge tubes in aliquots, label clearly and store at −20°C until required (see Note 6).

3.3 Peptide purification

1. Activation solution: 50% ACN in LC-MS grade water.

2. Equilibration solution: 0.5% TFA in 5% ACN.

3. Wash solution: 0.5% TFA in 5% ACN.

4. Elution buffer: 70% ACN in LC-MS grade water.

4. Step-by-step method details

4.1 Sample preparation

1. Collection of muscle tissue exudate (see Note 7): Take three 8 g cores (12 mm diameter × 2.5 cm) of muscle the muscle tissue under study (see Note 8).

2. Place the samples core in the spinning centrifuge tubes and centrifuge at 5911 ×g at 4°C in the ultracentrifuge for 60 min (see Note 9).
3. After centrifugation, collect the exudate in microcentrifuge tubes, mix well and spin down, make aliquots, label clearly, snap freeze in liquid nitrogen, and store at −80°C until required.
4. Determine the protein concentration in triplicate (e.g., Bradford, BCA assays) as per manufacturer's instructions.
5. Normalize the samples to a protein concentration of 100 µg in 100 µL total volume with lysis buffer in microcentrifuge tubes and freeze at −80°C until required (see Note 10).

4.2 Filter-aided sample preparation

1. Defrost the normalized samples on ice, mix well, spin down, and leave the normalized samples on ice until required.
2. Add 1 µL of DTT from a 0.5 M stock to the normalized samples and incubate for 20 min at 56°C using a heat block to denature the proteins by reducing disulfide bonds. Allow to fully cool down.
3. Prerinse the device (Amicon Ultra-0.5 Centrifugal Filter Devices) with buffer or LC-MS grade water to remove material that may interferes with the analysis (about 100 µL). If interference continues, rinse with 0.1 N NaOH followed by a second spin of buffer or LC-MS grade water (see Note 11).
4. Mix the 100 µg of protein lysate with 200 µL of pure 8 M Urea (made in LC-MS grade water).
5. Transfer the solution to the device (Amicon Ultra-0.5 Centrifugal Filter Devices) and spin using a microcentrifuge set to 14,000 ×g at 20°C for 40 min.
6. Discard the filtrate (flowthrough) and dilute the concentrate (top unit) with 200 µL of Wash buffer 11 and spin using a microcentrifuge set at 14,000 ×g at 20°C for 40 min.
7. Discard the filtrate and alkylate the concentrate with 100 µL of the alkylation solution.
8. Shake at 600 rpm for 1 min then incubate for 30 min in the dark at room temperature without mixing.
9. Centrifuge the device at 14,000 ×g at 20°C for 40 min.
10. Discard the filtrate and dilute the concentrate with 100 µL of Wash buffer 22 and centrifuge at 14,000 ×g at 20°C for 40 min.
11. Repeat step 10.

4.3 Protein digestion

1. Dilute the concentrate with 100 µL of freshly prepared 50 mM ammonium bicarbonate.
2. Add 2 µg of sequence-grade trypsin (1:50 enzyme:protein) (see Note 12).
3. Add 1 µL of 1% ProteaseMax Surfactant (see Note 13).

4. Transfer the filter unit to a new collection tube (see Note 14).
5. Digest the samples at 37°C overnight shaking at 300 rpm (see Note 15).
6. Next morning, centrifuge at 14,000 ×g at 20°C for 40 min. Retain the filtrate in the collection tube as this contains the peptides under study.
7. Rinse the filter unit with 50 μL of the 0.5 M NaCl rinse solution and spin at 14,000 ×g at 20°C for 20 min (see Note 16).
8. Remove the filter unit and discard. Acidify the sample to a final concentration of 2% ACN and 0.1% TFA to stop the digestion. Mix well.
9. Place samples at 37°C for 15 min and then centrifuge at 14,000 ×g for 5 min.
10. Store the filtrate that contains the peptides at −80°C until required or proceed with peptides purification.

4.4 Peptide purification

1. Tap column to settle the resin. Remove the top and bottom caps of the Pierce C18 Spin Columns (see Note 17) and place the column into a collection tube.
2. Carefully add 200 μL of Activation solution on all sides of the column to ensure none of the beads are stuck to the sides of the column and to wet resin.
3. Centrifuge at 1500 ×g for 1 min. Discard flowthrough.
4. Repeat steps 2 and 3.
5. Add 200 μL of equilibration solution and centrifuge at 1500 ×g for 1 min. Discard flowthrough.
6. Repeat step 5.
7. Load the peptide sample on top of the resin bed.
8. Centrifuge at 1500 ×g for 1 min.
9. Recover the flowthrough and repeat steps 7 and 8 two more times to allow the complete binding of peptides to the C18 resin (see Note 18).
10. Add 200 μL of Wash solution to the column.
11. Centrifuge at 1500 ×g for 1 min. Discard the flowthrough.
12. Repeat step 11 (see Note 19).
13. Place the column in a new collection tube.
14. Add 50 μL of Elution buffer to the top of the resin bed.
15. Centrifuge at 1500 ×g for 1 min.
16. Repeat steps 14 and 15 (see Note 20).
17. Discard the C18 column and gently dry the samples in a vacuum evaporator.
18. Freeze the peptide samples at −80°C until required.

4.5 Sample preparation for LC-MS analysis

1. Suspend the dried peptides in 50 μL solution of 2% ACN and 0.1% formic acid, vortex well and allow to stand at room temperature for 20 min.
2. Make two aliquots of 25 μL and store at −80°C until required or one aliquot can be directly analyzed by LC-MS.

5. Optimization and troubleshooting/notes

Note 1: Always weigh and then add SDS in the solution in the fume hood wearing a mask. It may be necessary to turn off the fume hood while weighing the SDS; however, as soon as the SDS is in solution the fume hood should be switched on to remove the SDS.

Note 2: Never heat urea solution above 37°C to avoid protein carbamylation. Do not freeze any thawed buffers containing urea. Make up aliquots, freeze and thaw aliquots as required (discard any remaining thawed material).

Note 3: This solution must be prepared freshly and used within a day.

Note 4: The precise blend of Trizma HCl and Trizma Base will bring the solutions to the desired pH (8.5 or 7.9); however, this must be confirmed with litmus paper.

Note 5: Use sequencing-grade or mass spectrometry grade trypsin (e.g., Promega's Trypsin Gold Mass Spectrometry grade). Sequencing grade modified trypsin reduces auto digestion, which may otherwise result in additional peptide fragments in the samples, which could interfere with downstream database searching of fragmented peptide masses.

Note 6: Keep the solution and aliquots on ice until ready for use.

Note 7: A simple sample preparation is pivotal to ensure a greater reproducibility. The specimen used in our studies is an exudate collected postmortem from muscle tissue following centrifugation (centrifugal drip [4]).

Note 8: Take care to avoid obvious pieces of fat.

Note 9: The amount of exudate collected using this method (three 8 g core samples, 5911 ×g, 4 °C for 60 min) ranged from about 80 μL to 2 mL (average of about 500 μL) [4]. It is possible to reduce the centrifugation speed (e.g., if it is not available, use an ultracentrifuge that can reach 6000 ×g) and increase the weight of the samples.

Note 10: Protein normalization can be performed the day before to perform FASP. It is recommended to randomize the samples preparation to reduce the batch effect.

Note 11: Do not allow the membrane to dry out once wet.

Note 12: The type of trypsin used may influence the ratio of enzyme:protein; see manufacturer's instructions.

Note 13: This step is optional but recommended to ensure efficient protein digestion with trypsin during the 3 h digestion period.

Note 14: This step can also be done before adding the trypsin.

Note 15: It is recommended to secure the lids of the device (Amicon Ultra-0.5 Centrifugal Filter Devices) using Parafilm wrap to avoid loss of samples due to evaporation and/or accidental opening of the lid.

Note 16: The device (Amicon Ultra-0.5 Centrifugal Filter Devices) must be rinsed post digestion to elute residual peptides that may be bound to walls of the device or the filter membrane.

Note 17: In our studies we used Pierce C18 Spin Columns (Thermo Fisher Scientific). Other devices are available, see manufacturer's instructions for methods.

Note 18: Flowthrough may be retained to confirm sample binding.

Note 19: If samples contain high levels of contaminants, repeat the wash step one to two additional times.

Note 20: This elution contains your eluted peptides. Do not discard.

References

[1] Auerbach D, Thaminy S, Hottiger MO, Stagljar I. The post-genomic era of interactive proteomics: facts and perspectives. Proteomics 2002;2:611.

[2] Lottspeich F. Proteome analysis: a pathway to the functional analysis of proteins. Angew Chem Int Ed 1999;38:2476.

[3] Martinez I, Friis TJ. Application of proteome analysis to seafood authentication. Proteomics 2004;4:347.

[4] Di Luca A, Mullen AM, Elia G, Davey G, Hamill RM. Centrifugal drip is an accessible source for protein indicators of pork ageing and water-holding capacity. Meat Sci 2011; 88:261–70. https://doi.org/10.1016/j.meatsci.2010.12.033.

[5] Di Luca A, Hamill R, Mullen AM, Elia G. Dige analysis of animal tissues. Diff Gel Electrophor 2018;1664. https://doi.org/10.1007/978-1-4939-7268-5_12.

[6] Abdallah C, Dumas-Gaudot E, Renaut J, Sergeant K. Gel-based and gel-free quantitative proteomics approaches at a glance. Int J Plant Genom 2012;2012:494572. https://doi.org/10.1155/2012/494572.

[7] Neilson KA, Ali NA, Muralidharan S, Mirzaei M, Mariani M, Assadourian G, et al. Less label, more free: approaches in label-free quantitative mass spectrometry. Proteomics 2011;11:535. https://doi.org/10.1002/pmic.201000553.

[8] Soderblom EJ, Philipp MF, Fau TJW, Fau CMG, Moseley MA. Quantitative label-free phosphoproteomics strategy for multifaceted experimental designs. Anal Chem JID 2011. 0370536.

[9] Bovo S, Di Luca A, Galimberti G, Dall'Olio S, Fontanesi L. A comparative analysis of label-free liquid chromatography-mass spectrometry liver proteomic profiles highlights metabolic differences between pig breeds. PLoS One 2018;13:1. https://doi.org/10.1371/journal.pone.0199649.

[10] Di Luca A, Ianni A, Henry M, Martino C, Meleady P, Martino G, et al. Label-free quantitative proteomics and stress responses in pigs—the case of short or long road transportation. PLoS One 2022;17(11):e0277950. https://doi.org/10.1371/journal.pone.0277950.

[11] Wisniewski JR, Zougman A, Nagaraj N, Mann M. Universal sample preparation method for proteome analysis. Nat Methods 2009;6:359–62. http://www.nature.com/nmeth/journal/v6/n5/suppinfo/nmeth.1322_S1.html.

Protein digestion

In-gel tryptic digestion approaches for protein and proteome characterization

Michael Henry[1] and Paula Meleady[1,2]

[1]*National Institute for Cellular Biotechnology, Dublin City University, Glasnevin, Dublin, Ireland;*
[2]*School of Biotechnology, Dublin City University, Glasnevin, Dublin, Ireland*

1. Introduction

In-gel digestion coupled with liquid chromatography (LC) and mass spectrometry (MS) is a very powerful tool when it comes to the identification and characterization of proteins. The identification of proteins from polyacrylamide gels can offer a number of advantages when compared to gel-free proteomic approaches especially for highly complex samples. The main advantages of in-gel digestion from SDS-PAGE separation are information on the molecular weight ranges from each gel slice/spot prepared, the complete protein solubilization using SDS and denaturation of the protein sample. Gel electrophoresis removes detergents, salts, and buffers which can interfere with digestion steps along with binding, elution, and ionization steps in LC-MS/MS analysis. Gel electrophoresis followed by total protein staining can act as an important sample quality control check prior to sample preparation for LC-MS/MS providing information on sample quality, complexity, abundance, etc.

In-gel digestion of proteins isolated by gel electrophoresis is a cornerstone of mass spectrometry—driven proteomics since the 1990s [1]. The original 16-year-old method by Shevchenko et al. [2] has been the go-to for in-gel digestion of proteins isolated by gel electrophoresis and mass spectrometry proteomics. In-gel digestion and LC-MS/MS from co-immunoprecipitation studies facilitate the study of protein—protein interaction networks to identify protein partners and provide comprehensive insight into the molecular pathways [3]. In-gel digestion also remains essential in workflows to identify and quantify by mass spectrometry the components of protein complexes fractionated by native PAGE [4]. Affinity purification—mass spectrometry is still the method of choice for discovering protein—protein-interactions under native condition [5].

The rapid method for in-gel digestion, described in this chapter, following Coomassie staining of a one-dimensional SDS-PAGE gel and LC-MS/MS analysis can be applied to multiple regions of a gel or when focusing on a specific region to target

Proteomics Mass Spectrometry Methods. https://doi.org/10.1016/B978-0-323-90395-0.00003-6

protein(s) of interest. This method can then be used in conjunction with a targeted MS approach to improve protein identification and protein sequence coverage. This chapter describes the use of a spent culture medium from a CHO cell line producing recombinant glycoprotein erythropoietin (EPO) [6] separated using SDS-PAGE followed by in-gel digestion, and a combination of a data-dependent acquisition method and a targeted mass spectrometry method for protein identification.

2. Materials and equipment

- Mass Spectrometry Grade Trypsin (Fisher Scientific 15255753)
- ProteaseMAX Surfactant, Trypsin Enhancer (Promega V2071). See Note 1
- Ammonium Bicarbonate BioUltra, ≥99.5% (Sigma−Aldrich 09830)
- DL-Dithiothreitol (DTT) BioUltra (Sigma−Aldrich 43815)
- Iodoacetamide BioUltra (Sigma−Aldrich I1149)
- Trifluoroacetic Acid (TFA), Optima LC/MS Grade, Fisher Chemical (Fisher Scientific 10723857)
- Formic Acid (FA), 99.0+%, Optima LC/MS Grade, Fisher Chemical (Fisher Scientific 10780320)
- Water, Optima LC/MS Grade, Fisher Chemical (Fisher Scientific 10728098)
- Acetonitrile (ACN), Optima LC/MS Grade, Fisher Chemical (Fisher Scientific 10489553)
- Methanol, Optima LC/MS Grade, Thermo Scientific (Fisher Scientific 10031094)
- Acetic Acid, Optima LC/MS Grade, Fisher Chemical (Fisher Scientific11377540)
- Brilliant Blue G-Colloidal 1× (SigmaAldrich B2025)
- Air-circulation thermostat (37°C and 75°C)
- Bench-top centrifuge
- Vacuum centrifuge
- Fumehood
- Micropipettes
- Top pan weighing balance
- Sonicating water bath
- Low Protein Binding Microcentrifuge Tubes (Fisher Scientific 90410)
- SpeedVac Vacuum dryer
- C18 Zip Tips with 0.6 μL bed (Merck ZTC18S008). See Note 2
- A nanoflow ultra-high performance liquid chromatography (UHPLC) instrument: e.g., Thermo Scientific UltiMate 3000 RSLCnano System. See Note 3
- High resolution, high sensitivity, fast Scanning MS instrument: e.g., Thermo Scientific Orbitrap Fusion Tribrid Mass Spectrometer. See Note 4
- Peptide trapping Column: Acclaim PepMap100, C18, 300 μm × 5 mm (Fisher Scientific 160454)

- Peptide resolving Column: Acclaim PepMap 100, 75 μm × 50 cm, 3 μm bead diameter column (Fisher Scientific 164570)
- MS database search program and protein sequence database: SEQUEST HT using NCBI and Swiss-Prot publicly available protein fasta sequence databases. See Note 5

3. Before you begin

The following buffers need to be made up.

1. Gel fixing solution (7% acetic acid)
   - 10 mL per gel: In a fumehood add 0.7 mL of acetic acid to 9.3 mL of water.
2. Gel destaining solution (40% methanol)
   - 10 mL per gel: Add 4 mL of methanol to 6 mL of water.
3. Ammonium bicarbonate (NH₄HCO₃) MW 79.06 g/L
   - Prepare 50 mL of 100 mM solution: Add 0.395 g and bring to 50 mL with LC-MS water.
4. 50 mM acetic acid solution for trypsin resuspension
   - Add 286 μL of glacial acetic and bring to a final volume of 10 mL in water.
5. Gel piece de-staining solution
   - Methanol: 100 mM Ammonium Bicarbonate (1:1, vol/vol).
6. Gel piece dehydration solution
   - Acetonitrile (ACN): 100 mM ammonium bicarbonate (1:1, vol/vol).
7. Protein reducing with Dithiothreitol—DTT—MW 154.25 g/L
   - Dissolve 0.154 g in 1 mL of 50 mM ammonium bicarbonate.
8. Protein alkylating with Iodoacetamide (C_2H_4INO) MW 184.86 g/L
   - Dissolve 0.102 g in 1 mL of 50 mM ammonium bicarbonate.
9. ProteaseMAX surfactant and trypsin enhancer
   - Add 100 μL of 100 mM ammonium bicarbonate to a new vial of ProteaseMAX surfactant for a 1% solution.
10. Trypsin—Thermo Scientific Pierce Trypsin Protease, MS Grade
    - Resuspend lyophilized trypsin using 50 mM acetic acid to a final concentration of 1 μg/μL.
11. C18 Zip Tip Buffers
    - Wetting solution prepare 10 mL. Add 7 mL of ACN and bring to a final volume of 10 mL with LC-MS grade water.
    - Prepare 10 mL of Equilibration and Washing Solution containing 0.1% TFA with 2% ACN. In a fumehood, add 10 μL of TFA to 200 μL of ACN and bring to a final volume of 10 mL with LC-MS grade water.
    - Elution solution: Prepare 10 mL in a fumehood, add 0.1 mL of TFA to 7 mL of ACN and bring to a final volume of 10 mL with LC-MS grade water.

4. Step-by-step method details

4.1 Processing of gel

Twenty microliters of spent culture medium from CHO cells producing recombinant EPO is separated using an SDS-PAGE gel. Following loading and running of the protein sample on an SDS-PAGE gel, the gel is fixed with 7% acetic acid in 40% (v/v) methanol for 30 min. The gel is then stained overnight with Brilliant Blue G— Colloidal 1X as per manufacturer's recommendations. The gel is destained with a 40% methanol solution until clear visualization of protein stained bands is observed (Fig. 9.1). The gel is then stored in LC-MS grade water.

4.2 In-gel digestion

1. Excise a small spot from the gel at the appropriate MW region using a sterile pasteur pipette and transfer this gel plug to a clean microcentrifuge tube.

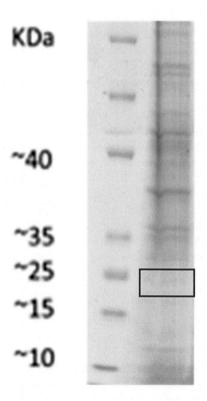

FIGURE 9.1

Stained gel using spent culture medium from Chinese hamster ovary (CHO) cells producing recombinant erythropoietin (EPO). The region of interest is highlighted in *red*.

2. Wash the gel piece with 100 µL of LC-MS water and vortex for 10 s. Remove and discard the solution.

3. Destain the gel piece with 100 µL of destaining solution, vortex for 10 s. Remove and discard the solution. Repeat at least once or until the stain has been removed.

4. Dehydrate the gel piece with 100 µL of dehydration solution for 5 min with occasional vortexing. Remove and discard the solution.

5. Dry the gel piece using a Speed Vac vacuum centrifuge until the gel piece is dry.

6. Rehydrate the gel piece and reduce the proteins using 100 µL of a freshly prepared 25 mM DTT solution prepared in 50 mM ammonium bicarbonate for 15 min at 75°C using an air-circulation thermostat. Remove and discard the solution. Allow the gel piece to cool to room temperature. See Note 6.

7. Alkylate the proteins in the gel piece using 100 µL of a freshly prepared 55 mM iodoacetamide solution prepared in 50 mM ammonium bicarbonate for 20 min in the dark at room temperature.

8. Dehydrate the band again with 100 µL of dehydration solution for 5 min with occasional vortexing. Remove and discard the solution.

9. Dry the gel piece using a Speed Vac vacuum centrifuge until the gel piece is dry.

10. Rehydrate the gel piece and digest using 20 µL of 12.5 ng/µL trypsin prepared in 0.01% ProteaseMAX surfactant in 50 mM ammonium bicarbonate for 10 min. Overlay with 30 µL of 50 mM ammonium bicarbonate.

11. Digest for 1 h at 50°C or 2–3 h at 37°C.

12. Following digestion, collect the condensate from the tube by centrifuging at 12,000 ×g for 10 s. Transfer the extracted peptides solution to a fresh tube and stop the enzymatic reaction by adding TFA to a final concentration of 0.5%. Centrifuge again at 12,000 ×g for 10 min as this step will minimize the potential introduction of non-peptide material into the LC-MS system and will also reduce the level of degraded ProteaseMax surfactant.

13. The digest can be analyzed directly by LC-MS or stored short term at −20°C.

4.3 Peptide desalting and concentration with zip-tip (optional)

If peptide samples require desalting and sample concentration, use a 10 µL Zip Tip (0.6 µL bed volume). See Note 7 and Note 8.

1. Slowly aspirate 20 µL of Wetting Solution into a new Zip Tip and discard to waste.

2. Equilibrate the Zip Tip by slowly aspirating 20 µL of Equilibration/Washing Solution and discard to waste. Repeat five times to fully equilibrate the Zip Tip.

3. Bind the peptide sample by aspirating and dispensing the sample with five cycles, and dispense finally to waste.

4. Wash peptide sample on the Zip Tip by slowly aspirating 20 µL of Equilibration/Washing solution to waste and repeat five times.

5. Elute peptides: Using the Elution solution, aspirate and dispense the peptide sample through the Zip Tip into a 10 µL clean microcentrifuge tube and repeat once.

6. Completely dry the tryptic peptides using a Speed Vac vacuum centrifuge
7. The digest can be analyzed directly by LC-MS or can be stored short term at −20°C.
8. For LC-MS analysis, resuspend samples in 10 μL of a solution containing 2% (v/v) ACN, 0.1% (v/v) TFA followed by sonication in a water bath for 5 min.

4.4 LC-MS/MS conditions for peptide identification

Use an UltiMate 3000 nano RSLC (Thermo Scientific) (or alternative system) to perform nano-flow reverse-phased capillary high-pressure liquid chromatography in combination with an Orbitrap Fusion Tribrid Mass Spectrometer (MS) (Thermo Scientific) (or alternative system) for peptide identification following in-gel digestion.

1. A 1 μL peptide sample is loaded onto the trapping column (PepMap100, C18, 5 mm) at a flow rate of 25 μL/min with 2% (v/v) ACN, 0.1% (v/v) TFA for 3 min.
2. The sample is then resolved onto an analytical column (Acclaim PepMap100, 75 μm × 50 cm) where a binary gradient is employed to elute the peptides over 60 min at 300 nL/min (LC running conditions are provided in Fig. 9.2).
3. Eluting peptides from the HPLC are analyzed by MS either using a data-dependent acquisition method or a targeted method where known peptides masses and their respective retention times are only targeted for MS/MS analysis.
4. In both methods, a voltage of 1.9 kV is applied to the stainless-steel emitter tip. Full scans within a 380–1500 m/z range are carried out in the Orbitrap mass analyzer at a resolution of 120,000 (at m/z 200), automatic gain control (AGC) target value of 4×10^5, and a maximum ion injection time of 50 ms (MS conditions is provided in Fig. 9.3).
5. For MS/MS scanning data-dependent acquisition (DDA), isolate the peptides in the instrument's Quadrupole using an isolation width of 1.6 m/z and fragment

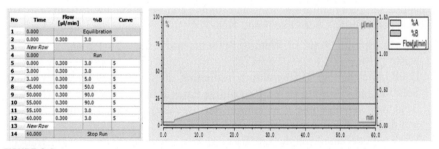

FIGURE 9.2

LC conditions for peptide separation. Buffer A consists of 100% water with 0.1% (v/v) FA and Buffer B consists of 80% ACN with 0.08% (v/v) FA.

FIGURE 9.3

MS and MS/MS conditions for peptide analysis following HPLC separation.

the peptides using Higher Energy Collision Dissociation (HCD) set at 28%. The ions generated with measured in the Orbitrap at a resolution of 30,000 with AGC target value of 5×10^4 and a maximum ion injection time of 300 ms (MS/MS conditions is provided in Fig. 9.3). See Note 9.

6. For data-dependent acquisition, use the top-speed acquisition algorithm and apply a dynamic exclusion after 60 s.
7. For targeted MS analysis use a target list with respective peptide masses, charge and expected retention times. A repeat count of 5 MS/MS's per target using a dynamic exclusion of 30 s can be employed (targeted MS method conditions are provided in Fig. 9.4). See Note 10.

FIGURE 9.4

Targeted MS method including peptides masses, charge, and expected retention times.

8. For data analysis, use the Thermo Scientific Proteome Discoverer 2.2 software using the SEQUEST HT search engine. Set the precursor mass tolerance to 10 ppm and fragment mass tolerance to 0.02 Da. Set the following parameters: Carbamidomethylation (+57.021 Da) for cysteine as a fixed modification and methionine oxidation (+15.996 Da) as a variable modification. Search the data against an NCBI *Cricetulus griseus* database and Swiss-Prot human database with a 1% FDR criteria using Percolator. See Note 11.

5. Anticipated results

As an example, the peptides from the band of interest were analyzed by LC-MS/MS using a data-dependent acquisition (DDA) method and a target m/z method.

The DDA method when analyzed using SEQUST HT in Proteome Discoverer resulted in an acquisition of over 4800 MS/MS scans from a 60 min HPLC separation which resulted in over 2300 confident peptide spectral matches from 191 *Cricetulus griseus* proteins and 1 *Homo sapiens* protein. The human protein Erythropoietin matched 6 peptides with 26% sequence coverage. A section of the overall protein identification results is provided in Fig. 9.5.

The targeted method when analyzed using SEQUST HT in Proteome Discoverer resulted in an acquisition of 267 ms/ms targeted scans over the 60 min HPLC separation, which resulted in 255 confident peptides spectral matches from 1 *Homo sapiens* protein Erythropoietin which had matched 9 peptides and with 47% sequence coverage. The protein identification result is provided in Fig. 9.6.

Protein FDR Confidence ⊕	Accession	Description	Coverage [%]	# Peptides	# PSMs	# Unique Peptides	MW [kDa] ▲
High	354501786	cofilin-1 [Cricetulus griseus]	45%	7	22	7	18.5
High	1868092478	prostaglandin E synthase 3 isoform X2 [Cricetulus griseus]	23%	3	5	3	19.6
High	625204337	60S ribosomal protein L11 [Cricetulus griseus]	25%	3	3	3	20.3
High	P01588	Erythropoietin OS=Homo sapiens OX=9606 GN=EPO PE=1 SV=1	26%	6	39	6	21.3
High	1537982271	flavin reductase (NADPH) isoform X2 [Cricetulus griseus]	40%	5	13	5	22.2
High	350537945	peroxiredoxin-1 [Cricetulus griseus]	73%	13	30	13	22.2
High	354500404	metalloproteinase inhibitor 1 [Cricetulus griseus]	47%	8	14	8	22.4
High	1538026942	clathrin light chain B isoform X2 [Cricetulus griseus]	9%	2	2	2	23.1
High	1868064058	nascent polypeptide-associated complex subunit alpha [Cricetulus grise	13%	2	2	2	23.4
High	354469007	rho GDP-dissociation inhibitor 1 [Cricetulus griseus]	15%	2	2	2	23.4
High	354474350	peptidyl-prolyl cis-trans isomerase B [Cricetulus griseus]	12%	2	2	2	23.6
High	350537543	glutathione S-transferase P 1 [Cricetulus griseus]	60%	7	18	7	23.6
High	354470166	40S ribosomal protein S8 [Cricetulus griseus]	12%	2	2	2	24.2

FIGURE 9.5

A section from the protein results using Proteome Discoverer following DDA analysis of proteins identified and in-gel digestion. In total, 192 high confident proteins were identified with a minimum of 2 peptides. They included 191 *Cricetulus griseus* proteins and 1 erythropoietin protein from *Homo sapiens*.

Protein FDR Confidence ⊞	Accession	Description	▲	Coverage [%]	# Peptides	# PSMs	# Unique Peptides	MW [kDa]
High	P01588	Erythropoietin OS=Homo sapiens OX=9606 GN=EPO PE=1 SV=1		47%	9	255	9	21.3

FIGURE 9.6

One protein was identified using the targeted method. Erythropoietin protein from *Homo sapiens* identified by proteome discoverer with 9 unique peptides following targeted MS analysis of peptides following in-gel digestion.

6. Optimization and troubleshooting/notes

Note 1. Although we recommend the use of ProteaseMax surfactant it can be left out of the digestion step. We would then recommend using 50 μL of trypsin solution to ensure the gel piece is completely covered with solution and it will not dry out and we would also recommend overnight digestion at 37°C.

Note 2. Avoid micro-C18 (0.2 μL bed volume) tips for peptide desalt, concentration and purification step. The use of micro-C18 tips can result in a decrease in peptide recovery due to competition of the degraded surfactant and peptides for binding sites.

Note 3. UHPLC for peptides analysis allows for increased resolution and speed of peptide separations prior to MS detection.

Note 4. MS-based proteomics and peptide identification by tandem mass spectrometry is the dominant proteomics workflow for protein characterization.

Note 5. Search engines match peptide sequences with tandem mass spectra produced by the mass spectrometer to identify proteins and use protein sequence databases to suggest peptide candidates for consideration.

Note 6. Although proteins from gel electrophoresis have already been reduced, it is recommended that the additional reduction/alkylation step is performed. However, if rapid identification is preferred then this step can be skipped and you can proceed to digestion following the dehydration step after destaining.

Note 7. If the peptide signal is expected to be low, the peptides can be concentrated with a 10 μL (0.6 μL bed resin volume) C18 Zip Tip (Millipore). We would suggest to avoid using the micro-C18 (0.2 μL bed volume) tips for peptide cleanup as this small bed volume can result in a decrease in peptide recovery due to competition of the degraded surfactant and peptides for binding sites.

Note 8. Zip Tips and their C18 resin bed can produce some back pressure, therefore always depress the micropipette plunger to a dead stop and slowly release or dispense at each step to avoid introducing air.

Note 9. MS conditions used are suitable for high resolution nanospray ionization mass spectrometry; however, manufacturer recommended settings using ion trap or matrix-assisted laser desorption ionization mass spectrometry instruments will also confidently identify prepared peptide samples.

Note 10. Targeted MS methods can improve sequence coverage of known protein(s). Actual precursor masses and LC retention times from previous mass spectrometry data can help build target lists; however, theoretical precursor mass information can also be used for targeted MS. Characterization studies should consider other enzymes to help improve sequence coverage.

Note 11. Protein identification used was the tandem mass spectrometry database search program SEQUEST HT; however, other search programs such as MASCOT, PEAKS, Andromeda, ProteinPilot, etc., can be used for protein identification using the manufactures recommended settings.

7. Safety considerations

The techniques described here use biological materials and chemicals. Personal protection equipment must be worn at all times. When working with chemicals you should be very familiar with their Material Safety Data Sheets ahead of use. Validated biological safety cabinets and fume hoods should only be used. All waste generated (biological and chemical) should be disposed of in accordance with local guidelines and procedures.

8. Summary

In this chapter, we have provided a methodology for in-gel digestion of protein samples and their analysis using state-of-the-art mass spectrometry. The techniques described can be applied to any biological sample that has been separated on an SDS-PAGE gel.

Acknowledgments

This work was supported by a Science Foundation Ireland (SFI) Frontiers for the Future Award (grant no. 19/FPP/6759).

References

[1] Goodman JK, Zampronio CG, Jones AME, Hernandez-Fernaud JR. Updates of the in-gel digestion method for protein analysis by mass spectrometry. Proteomics December 2018; 18(23):e1800236. https://doi.org/10.1002/pmic.201800236. Epub 2018 Nov 25. PMID: 30259661; PMCID: PMC6492177.

[2] Shevchenko A, Tomas H, Havlis J, Olsen JV, Mann M. In-gel digestion for mass spectrometric characterization of proteins and proteomes. Nat Protoc 2006;1(6):2856–60. https://doi.org/10.1038/nprot.2006.468.

[3] Zhang M, Zhang K, Wang J, Liu Y, Liu G, et al. Immunoprecipitation and mass spectrometry define TET1 interactome during oligodendrocyte differentiation. Cell Biosci 2020;10:110. https://doi.org/10.1186/s13578-020-00473-5.

[4] Swart C, Martínez-Jaime S, Gorka M, Zander K, Graf A. Hit-gel: streamlining in-gel protein digestion for high-throughput proteomics experiments. Sci Rep 2018;8:8582. https://doi.org/10.1038/s41598-018-26639-3.

[5] Zhang Y, Sun H, Zhang J, Brasier AR, Zhao Y. Quantitative assessment of the effects of trypsin digestion methods on affinity purification-mass spectrometry-based protein-protein interaction analysis. J Proteome Res August 4, 2017;16(8):3068−82. https://doi.org/10.1021/acs.jproteome.7b00432. Epub 2017 Jul 20. PMID: 28726418; PMCID: PMC5656008.

[6] Costello A, Lao NT, Barron N, Clynes M. Improved yield of rhEPO in CHO cells with synthetic 5' UTR. Biotechnol Lett February 2019;41(2):231−9. https://doi.org/10.1007/s10529-018-2632-2. Epub 2018 Nov 30. PMID: 30506229.

MS proteomic analysis protocols

Protocols for label-free LC-MS/MS proteomic analysis

10

Esen Efeoglu[1], Michael Henry[1] and Paula Meleady[1,2]

[1]*National Institute for Cellular Biotechnology, Dublin City University, Glasnevin, Dublin, Ireland;*
[2]*School of Biotechnology, Dublin City University, Glasnevin, Dublin, Ireland*

1. Introduction

Mass spectrometry—based proteomic analysis has become an important complementary approach to genomic techniques due to the advancements in instrumentation (high-resolution mass spectrometers), improved databases, novel sample preparation and enrichment strategies (subcellular fractionation, rapid sample preparation, and enrichment kits). Mass spectrometry—based proteomics has gathered further attention due to its capability to explore posttranslational modifications to which genomics techniques are blind [1—4]. The phosphorylated and ubiquitinated proteomes of biological samples have started to be investigated to identify cellular states such as cancer versus healthy as well as to determine cellular mechanisms as a driver of the bioproductivity of CHO cells in the production of recombinant proteins [5—10].

For quantitative mass spectrometry—based proteomics, label-free and labeled approaches can be used for the investigation of biological samples such as cells, tissues, and bodily fluids [11—17]. Labeled quantitative proteomic analysis is based on the introduction of an internal standard (isotopomer of an analyte) with known quantity and measurement of the peak intensities to determine the quantity of the analyte. Among quantitative approaches, Stable Isotope Labeling of Amino Acids in Culture (SILAC) utilizes metabolic labeling at the cellular level while cells are still in culture conditions. As an alternative to metabolic labeling, chemical labeling can be used to label the cellular proteome after it has been extracted from a biological sample. Tandem Mass Tags (TMT), Isotope Coded Affinity Tags (ICAT), and Isobaric Tags for Relative and Absolute Quantification (iTRAQ) are some of the commonly used labels for the quantitative labeled approach. Even though labeled quantification has various advantages such as ease of comparability of the samples and less machine time, their use increases the complexity of sample preparation and total cost. Moreover, the number of samples which can be measured in one run remains limited due to the number of available tags [18,19].

Label-free quantitative proteomic analysis allows the measurement of protein abundances across multiple biological samples and offers flexible study design, low cost, and ease of sample preparation, although requires longer machine time.

Proteomics Mass Spectrometry Methods. https://doi.org/10.1016/B978-0-323-90395-0.00012-7

Biological samples ideally with multiple replicates (minimum of 3) are prepared simultaneously (extraction of proteins, followed by digestion of protein samples to peptides) and analyzed under identical run conditions for comparison of samples to minimize technical variances. Quantitation can be carried out as label-free quant at the MS1 level (peptide level) or as spectral counting at the MS2 level (fragment ions). At the MS1 level, the comparison is carried out among the peak areas of the peptide peak pairs across the samples. At the MS2 level, the number of MS/MS measurements is compared for a peptide peak across LC-MS runs. Comparison of peak area intensities or MS/MS measurement across sample groups provides fold changes which represent the expression profiles of the proteins across the study group (Fig. 10.1).

Both labeled and label-free quantification produce large and complex data sets which require computational analysis of the data to determine variances in the expression profiles across the sample groups. This has led to the development of bio-informatics software programs such as Progenesis QI for proteomics and Maxquant LFQ [20,21]. MaxQuant (https://maxquant.org/) is a publicly available software which provides the analysis of large proteomics data sets. Since its release in 2008, it has grown substantially in functionality and become one of the most frequently used platforms for mass spectrometry—based proteomics data analysis. Maxquant LFQ uses maximum peptide ratio information extracted from peptide peak ion intensities to quantify proteins across the sample groups. Progenesis QI for Proteomics (Waters Company, Nonlinear Dynamics, www.nonlinear.com) also provides accurate quantification of each protein using peptide ion peak intensities. Each peptide ion peak intensity is represented as a black spot on a map with m/z

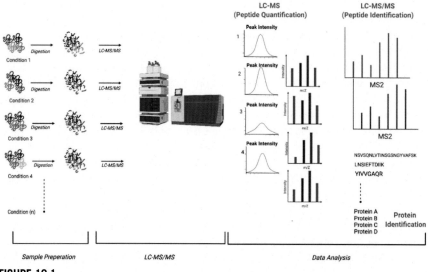

FIGURE 10.1

General workflow for the label-free LC-MS/MS proteomic analysis.

values plotted against retention time for each sample. The peak intensities across different runs are then aligned and normalized automatically (or alternatively manually) for comparison. Experimental design is set by the user to allow the identification of differentially expressed peptide ions within sample groups. The differentially expressed peptide ions are exported to identify proteins by using a database search engine (e.g., MASCOT, SEQUEST, and PEAKS Studio) and peptide ions which have been attributed to proteins are imported back into Progenesis QI for Proteomics. Peptides attributed to a single protein are accumulated and protein level abundances and fold changes are calculated.

In this chapter, we explain the common methodology that we use in our group for label-free quantitative analysis of biological samples. Parental (non-resistant) and chemotherapy drug-resistant lung cancer cells are used as a model biological sample to describe the sample preparation steps. Data acquisition parameters and data analysis steps using Progenesis QI for Proteomics for differential expression analysis and Proteome Discoverer as a search engine are also described.

2. Materials and equipment

- Nutrient Mixture F-12 (DMEM-F12) (Gibco, Thermo Fisher Scientific, Cat. No: 11320033)
- Fetal Bovine Serum (FBS) (Thermo Fisher Scientific, Cat. No:10270106)
- L-glutamine (Thermo Fisher Scientific, Cat. No:25030024)
- Trypsin (Gibco, Thermo Fisher Scientific, Cat No: 15400054)
- Mem-PER Plus Membrane Protein Extraction Kit (ThermoFisher Scientific, Cat. No: 89842)
- PreOmics iST kit (PreOmics GmBh, Germany; Cat No: P.O.00001)
- BCA Assay (Thermo Fisher Scientific, Cat. No: 23225)
- Acetonitrile Optima LC/MS grade (ACN) (Fisher Scientific, Ireland; Cat No: 10001334)
- Pierce Formic Acid LC-MS Grade (Thermo Fisher Scientific, Ireland; Cat No: 85178)
- MultiSKAN GO Plate reader (Thermo Fisher Scientific)
- Refrigerated centrifuge (capable of spinning 2 mL tubes at 3800 xg)
- Vortex
- SpeedVac Vacuum dryer
- Sonicator
- UltiMate 3000 RSLCnano System (Thermo Scientific)
- Orbitrap Fusion Tribrid Mass Spectrometer (Thermo Scientific)
- Proteome Discoverer Version 2.2 (Thermo Scientific)
- Progenesis QI for Proteomics (Waters, Nonlinear Dynamics)

3. Before you begin

This chapter focuses on the data acquisition and analysis of LC-MS/MS proteomic analysis. Although sample preparation techniques are out of the scope of this chapter, and are detailed in many of the chapters in this book, they are summarized and illustrated in Fig. 10.2. Further details regarding sample preparation can be found on the manufacturers' websites.

Samples used in study: Cancer cell line (Parental Cells) and its drug-resistant variant (Resistant Cells).

3.1 Summary of sample preparation steps

➤ Collection of cell pellets

Both parental and resistant cells are grown in T175 flasks in their respective growth medium (DMEM-F12) supplemented with 10% FBS and 4 mM L-Glutamine. Once cells reach 80%–90% confluence, cells are detached from the surface via trypsinization and centrifuged at 300× g for 5 min. Following centrifugation, the supernatant is removed and cells are washed with sterile PBS three times (Note 1).

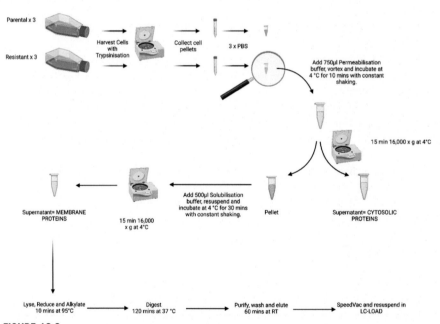

FIGURE 10.2

Schematic of sample preparation from parental and resistant lung cancer cells from culture to mass spectrometry.

➤ Membrane fractionation

Membrane and cytoplasmic fractions of the parental and resistant cells are separated using the MEM-PER Plus Protein extraction kit. The kit contains a cell wash solution, solubilization buffer and permeabilization buffer (Note 2). Solubilization buffer is used to lyse the cells to extract and collect the cytoplasmic proteins, whereas permeabilization buffer disintegrates membrane structures to extract membrane-related proteins and integral transmembrane proteins. The kit provides this separation between the cytoplasm and membrane-related proteins with 90% efficiency.

➤ Protein quantification

BCA assay is a commonly used protein detection assay which provides an accurate determination of protein concentration in various cell types such as affinity column fractions, purified proteins, industry-related contaminant proteins, cell lines, tissues, and bodily fluids.

➤ PreOmics iST kit

The PreOmics iST kit allows rapid and straightforward sample preparation for proteomic applications, and is outlined in detail in Chapter 3. The kit contains LYSE buffer for extraction of proteins from a biological sample which also reduces and alkylates the protein. The DIGEST contains a mixture of LysC and trypsin and effectively digests protein samples within 1–3 h at room temperature. The kit also includes solutions to wash (WASH1 and WASH2) and elute (ELUTE) the peptides. For liquid chromatography mass spectrometry LC-LOAD solution is used for the resuspension of dried peptides [22].

4. **Step-by-step method details**

4.1 **Mass spectrometry**

Several methods, data-dependent acquisition (DDA), data-independent acquisition (DIA), and targeted data acquisition (TDA), have been established and are commonly used for data acquisition via mass spectrometry. In this chapter, we focused on DDA which uses predetermined settings such as precursor intensity and charge state for fragmentation of a defined number of precursor ions from the full scan. DDA is used by our group routinely for the acquisition of MS data for the label-free quantitative analysis of biological samples. The liquid chromatography parameters, MS and MS/MS conditions, are shown in Tables 10.1 and 10.2, respectively.

By using the LC, MS, and MS/MS parameters outlined below, it is possible to obtain over 4000 protein IDs with a 35% conversion rate from MS/MS to PSMs from the proteome of a mammalian cell (Fig. 10.3).

Table 10.1 Liquid chromatography conditions.

Instrument	UltiMate 3000 RSLCnano System
Trapping column	PepMap100, C18, 300 μm × 5 mm, 25 μL/min
Resolving column	Acclaim PepMap 100, 75 μm × 50 cm, 3 μm bead diameter column, 300 nL/min
Column temperature	45°C
Trapping column-solvent and flow rate	25 μL/min, with 2% (v/v) acetonitrile (ACN), 0.1% (v/v) trifluoroacetic acid (TFA) for a 3 min desalting and concentration step
Resolving column—gradient	A binary gradient of solvent A (0.1% (v/v) formic acid in LC-MS grade water) and solvent B (80% (v/v) ACN, 0.08% (v/v) formic acid in LC-MS grade water) using 2%—32% B for 75 min, 32%—90% B in 5 min and holding at 90% for 5 min at a flow rate of 300 nL/min was used to elute peptides
Maximum loading amount	1 μg

Table 10.2 MS and MS/MS conditions.

MS instrument	Thermo Scientific Orbitrap Fusion Tribrid Mass Spectrometer
MS conditions	
Detector type	Orbitrap
Orbitrap resolution	120,000
Quadrupole isolation	✔
Scan range	380—1500
Cone voltage	1900
MS/MS conditions	
Isolation mode	Quadrupole
Isolation window (m/z)	1.6
Activation type	HCD
Collision energy (%)	28
Detector type	Ion trap
Maximum injection time (ms)	35

4.2 Analysis of label-free proteomics data

Proteome Discoverer is a node-based, effective, and accurate platform for the analysis of highly complex mass spectrometry—based proteomics data. Proteome

	Protein IDs	Peptides	PSMs	MS/MS	Success rate
Mammalian cell lines	~4000-5000	22000-28000	30000-35000	80000	%35

FIGURE 10.3

Number of protein IDs, peptides, PSMs, MS/MS count, and success rate achieved from the analysis of a mammalian whole cell lysate in Thermo Scientific Orbitrap Fusion Tribrid mass spectrometer coupled with UltiMate 3000 RSLCnano System using the parameters outlined in Tables 10.1 and 10.2. Data base search results (FDR <1%), Sequest HT 10 ppm Precursor/0.6 Da fragments.

Discoverer supports workflows for label-free quantitative analysis as well as isotopically labeled quantitative analysis such as TMT and SILAC labeling. In this chapter, label-free quantification of data sets obtained from parental and drug-resistant lung cancer are used to show the workflow for Proteome Discoverer 2.4 software. The user can follow the steps outlined below for the proteomic analysis of their data:

1. Open File > New Study, this opens the "New study and Analysis" window (Fig. 10.4).
2. In a "New study and Analysis" window, define: Study name, Study root directory, Processing workflow, Consensus workflow and add files. It is important to note that Proteome Discoverer accepts file types such as .raw, .mgf, msf, mzML, mzData, and mz XML. The user can choose any of these files to be uploaded into the Proteome Discoverer software.

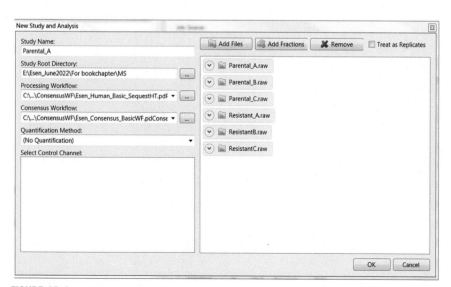

FIGURE 10.4

New study and analysis set up.

3. Click "OK" and the "Study" opens as a new tab. The added files can be found in the "Input Files" tab under the study window.

4. Set the nodes for processing and consensus workflows. Briefly,

Click on the Processing step in the analysis window and click on the "workflow" tab (Fig. 10.5).

An example workflow tree for processing and consensus is provided in Fig. 10.6. In the processing workflow, parameters for spectrum files such as protein database, enzyme name (e.g., trypsin or trypsin + LysC), and dynamic and static modifications can be set. The spectrum selector also provides a section to include several parameters such as minimum and maximum precursor mass, lowest and highest charge limits, activation type, and maximum collision energy. Sequest HT is a search algorithm and includes the selection of the following parameters: input data (protein database, enzyme name, maximum number of missed cleavages, and minimum and maximum peptide length), tolerances (precursor mass tolerance and fragment mass tolerance), spectrum matching (based on the selection of a, b, c, x, y, and z ions and their weights), as well as dynamic and static modifications for the search.

For consensus workflow, drag the nodes (shown in Fig. 10.5) from the left panel into the consensus workflow to obtain meaningful results from the analysis. Proteome Discoverer generates its output as MSF files and this node contains only the proteins, PSMs, and MS/MS spectrum information. It does not include confidence or scores. The PSM Grouper groups the PSMs into peptide groups. The validation methodology can be set in the Peptide Validator node. Further filtering of PSMs can be provided in the peptide and protein filtering node. The

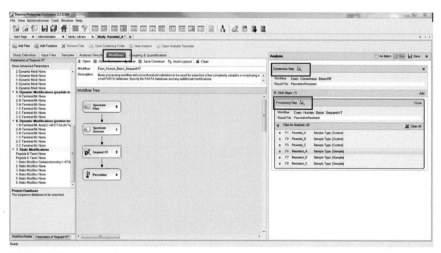

FIGURE 10.5

Setting up processing and consensus workflows.

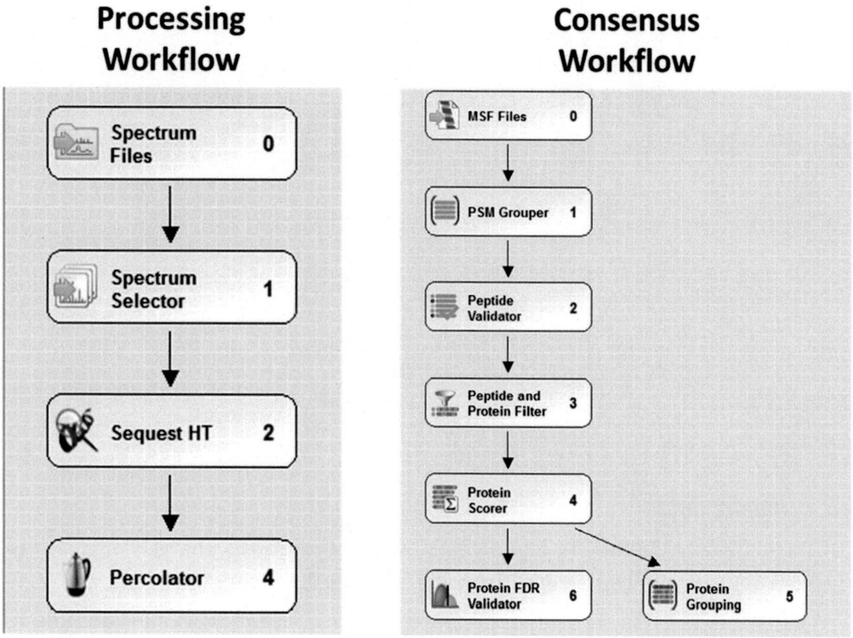

Processing Workflow

- Spectrum Files — 0
- Spectrum Selector — 1
- Sequest HT — 2
- Percolator — 4

Consensus Workflow

- MSF Files — 0
- PSM Grouper — 1
- Peptide Validator — 2
- Peptide and Protein Filter — 3
- Protein Scorer — 4
- Protein FDR Validator — 6
- Protein Grouping — 5

FIGURE 10.6

An example workflow tree for processing and consensus.

protein score of a protein is then calculated via the protein scorer node. The protein FDR Validator node calculates the FDRs for proteins.

5. Once processing and consensus workflows are set, click "Run" in the analysis window.

6. From the Job queue, double-click on the completed consensus run.

7. In the consensus file various parameters such as protein description, protein coverage, associated tables, number of peptides, PSMs, and unique peptides can be explored. It is also possible to set filters for filtering based on various parameters (e.g., keep only master proteins).

4.3 The use of Progenesis QI for Proteomics for the analysis of label-free proteomics data

The Progenesis QI for Proteomics software platform provides a qualitative and quantitative analysis of the label-free mass spectrometry—based proteomics data with consistency and precision. It is important to note that the Progenesis QI for Proteomics software allows the analysis of the labeled data (e.g., TMT labeling), however labeled analysis is not in the scope of this chapter. The workflow from Progenesis QI for Proteomics is outlined below and shown in Fig. 10.7.

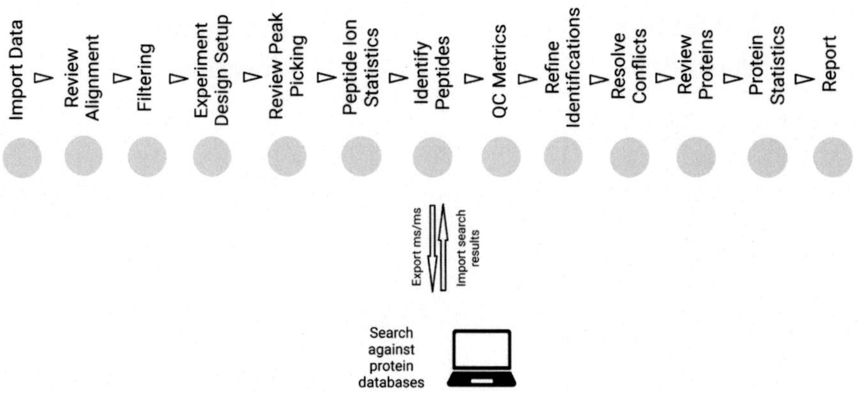

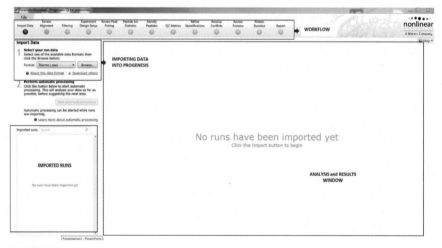

FIGURE 10.7

Workflow of Progenesis QI for Proteomics.

4.3.1 Import data

1. Create a folder containing raw data and where Progenesis analysis data will be saved after the analysis (e.g., E:\April2022\DrugResistance\RawData and \Progenesis).
2. Open the Progenesis QI for Proteomics and select File > Create new LC-MS experiment.
3. Name the label-free experiment (e.g., ParentalvsResistant), choose machine type (e.g., high-resolution mass spectrometer) and choose the experiment folder. (e.g., E:\April2022\DrugResistance\Progenesis). Click "Create experiment."
4. The main page for the Progenesis QI for Proteomics will open up showing workflow (Fig. 10.8).

FIGURE 10.8

Starting up Progenesis QI, workflow bar, and importing data.

5. Select your run data by browsing files (e.g., E:\April2022\DrugResistance\Raw data) and import the raw data into the Progenesis QI for Proteomics software (Fig. 10.8).
6. Once runs are imported, start automatic processing by following the steps outlined below:
7. Perform automatic processing > Start automatic processing > Assess all runs in the experiment for suitability > "Next" > Yes, automatically align my runs > "Next" > "Finish." (Notes 3 and 4). The software can objectively choose the best reference run and then apply alignment automatically. The quality of this process is reported by scores and visual displays (Note 5).

4.3.2 Review alignment

8. Once automatic processing is completed, press "Review Alignment" (Fig. 10.9)
9. Once the alignment of the runs is reviewed by the user, press "Section Complete" to start alignment by setting peak picking parameters.
10. Alignment is the next step to filter the data by setting peak picking parameters (Note 6).

    Peak picking parameters > Choose runs for peak picking: Select all the runs for peak peaking.

    Peak picking parameters > Peak peaking limits > Sensitivity: Select automatic.

    Peak picking parameters > Maximum allowable charge > Maximum ion charge = 20.

    Peak picking parameters > Retention time limits: no limits.

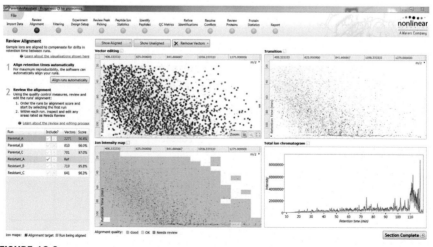

FIGURE 10.9

Alignment of runs.

4.3.3 Filtering

11. Once alignment is finished, the user can define normalization and filtering parameters or alternatively proceed to the "experimental design" without filtering the data (Note 7).
12. Click "Section Complete" once filtering parameters are set.

4.3.4 Experiment design setup

13. Define experiment design as either a between-subject design or a within-subject design (Fig. 10.10 and Note 8).
14. Click "Create" (for between-subject design), assign an experiment name (e.g., Parental vs. Resistant) and click "Create Design."
15. Set up the conditions on the left side panel of the screen, group related runs under the conditions and click "Section Complete."

4.3.5 Review peak picking

16. In the following step, Review Pick Picking, the user can define parameters to filter the peaks (e.g., ANOVA P-value and/or Q-value). Once parameters are set click "Create" on the left side panel to apply the set parameters to the data set by setting filters. If no parameters have been set, click "Section Complete."

4.3.6 Peptide ion statistics

17. Peptide ion statistics shows the principal component analysis of identified peptide ions and their standard expression profiles, once the review is completed, click "Section Complete."

(A)   **(B)**

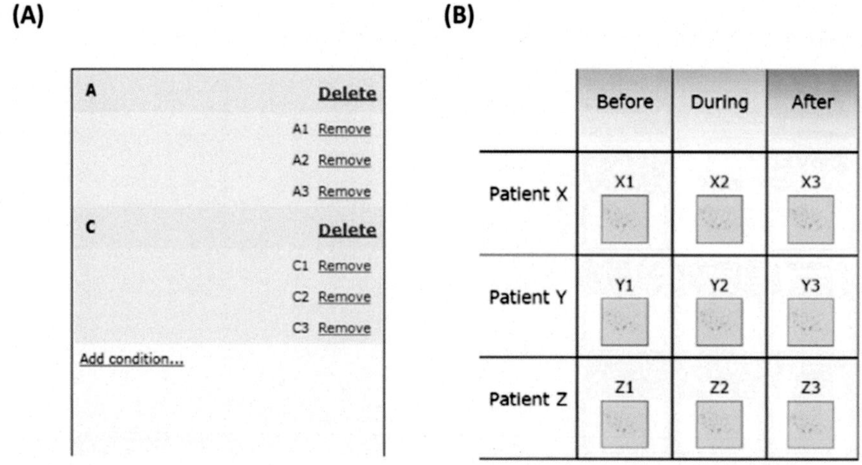

FIGURE 10.10

Experimental design in Progenesis QI for Proteomics. Between subject design (A) and within subject design (B).

4.3.7 Identify peptides

18. The identification of the peptides appears as the next step. From the bottom of the left panel, select the program of choice (e.g., Proteome Discoverer using SEQUEST) and click "Export (number of identified) ms/ms spectra." The MS/MS data here can be exported in a format of choice suitable for the search algorithm of choice for a sample such as a *.mgf (mascot generic file), and searched against the fasta database (human, mouse, Chinese hamster, etc.) using Proteome Discoverer and SEQUEST HT.

19. Search the .mgf file in Proteome Discoverer or the choice of another compatible search engine. The search of the .mgf file using SEQUEST results in the identification of peptides and associated proteins (Fig. 10.11). However, it is important to note that identified PSMs are imported back to Progenesis QI for peptide and protein matching; filtering of the PSMs at this stage based on scoring evaluation and the FDR values improves the quality of the data (Box 10.1).

20. Export the Proteome Discoverer results as .PEPXML file and import it back to Progenesis QI by clicking the "import search results" on the left side panel (Fig. 10.13).

4.3.8 QC metrics, refine identifications, and resolve conflicts

21. Once peptides are confidently identified, QC metrics appear in the next step and identifications can be refined further based on different parameters (e.g., score).

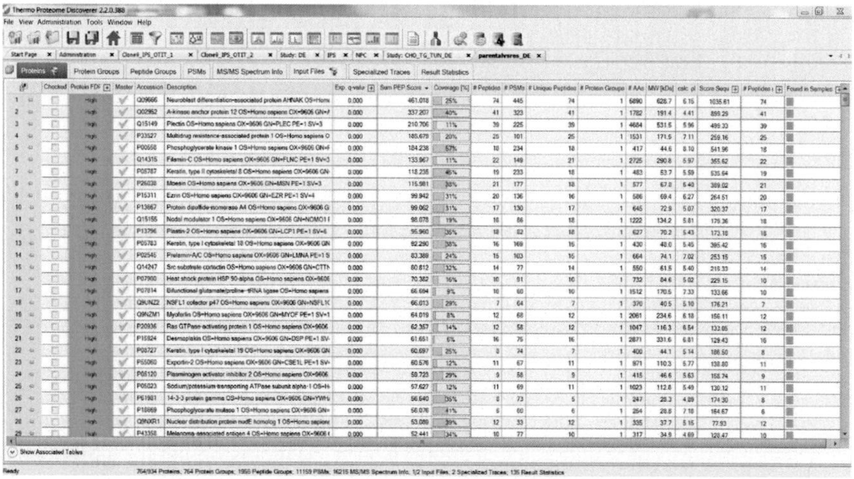

FIGURE 10.11

Differential expression data obtained from the SEQUEST search of exported data from Progenesis QI.

Box 10.1 An overall look at SEQUEST Search Data in Proteome Discoverer

The search data includes information related to proteins, peptide groups, and PSMs identified from the sample. Fig. 10.12 shows an outline of the information which can be gathered from the SEQUEST search of the differential Expression data.

For example, multidrug resistance—associated protein 1 (MRP-1) is identified in our differentially expressed data sets with 20% coverage of full sequence. In total 101 PSMs associated with 25 peptides are identified for this protein with a high FDR value. One of these peptides which contributed to the identification of MRP-1 was "GSVAYVPQQAWIQNDSLR" and it was detected by 3 MS/MS at elution time of 96.6 min and XCorr values of over 2.69 and above. The user can also visualize the Fragment Match spectrum for the identified peptide for quality control.

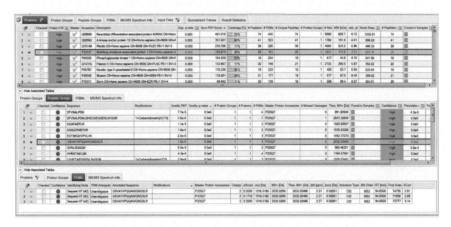

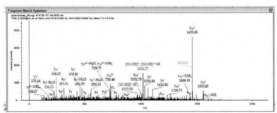

FIGURE 10.12

Example of SEQUEST search result obtained from DE analysis of parental and resistant lung cancer cells.

22. At this step, the user has options for any peptides/features that do not have enough MS/MS data and you can export a targeted inclusion list for a replicate run of samples to gather more MS/MS data to increase your identification coverage for the peptides of interest.

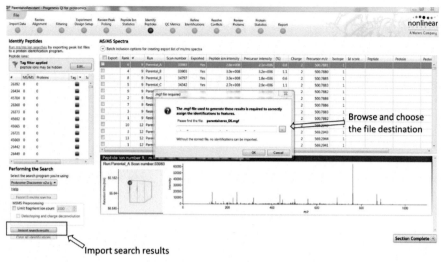

FIGURE 10.13

Importing the search results back to Progenesis QI.

23. The user can resolve conflicts which can arise from one peptide being assigned to more than one protein by using parameters such as the contribution of the peptide to the overall protein score, number of identified peptides, score value, etc. Once this is done, click "Section Complete."

4.3.9 Review proteins

24. In the next steps, the user can "Review Proteins," "Protein statistics," and export "Report" by using principal component analysis results and standard expression profiles. This step also allows setting more parameters for filtering of the results based on ANOVA value, number of identified peptides, maximum fold change, etc. (Fig. 10.14).

The Section 4.3.9 also allows the visualization of the peptides and their contributions to final protein identification. Fig. 10.15 shows the peptides and their normalized abundances in parental and resistant lung cancer cells.

In the Section 4.3.9, the user can export peptide and protein measurements to excel by clicking "Export protein measurements" and/or "Export peptide measurement" on the left side panel (4: Export data for further processing).

4.3.10 Protein statistics

25. The Section 4.3.10 provides an overview of the principal component analysis of the data, ANOVA (*P*) and *Q*-values of the identified proteins as well as the expression profiles of the proteins in parental and resistant cells (Fig. 10.16).

FIGURE 10.14

Creating filters for the results.

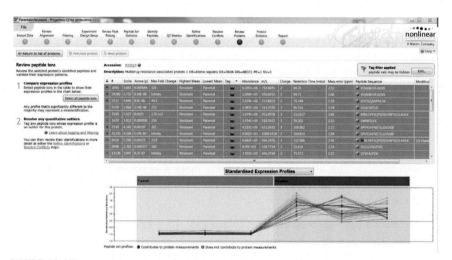

FIGURE 10.15

List of peptides which contribute to the identification of multidrug resistance-associated protein 1 (ABCC1) and expression profiles of these peptides in parental and resistant cells.

4.3.11 Report

26. In the Section 4.3.11, the user can export the data as a .pdf or HTML file including various parameters, such as peptide lists, statistics, abundances, protein lists, statistics, and abundances. Fig. 10.17 shows an example of this

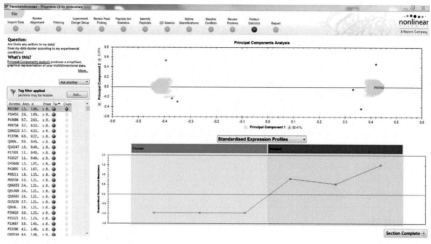

FIGURE 10.16

Protein statistics.

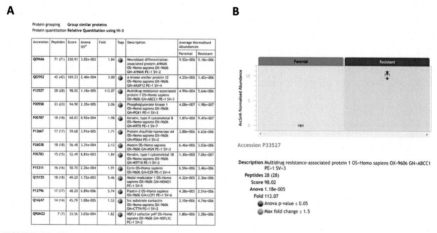

FIGURE 10.17

Example of list of proteins exported from the differential expression analysis of parental and resistant lung cancer cell lines (A) and plot of normalized abundances of ABCC1 in parental and resistant cell lines (B).

exported document including a list of proteins with their abundances, fold changes and ANOVA values (A) and a plot of the expression profile of multidrug resistance-associated protein 1 (ABCC1).

5. Safety considerations and standards

Protocols described in this chapter require the use of biological materials as well as various chemicals for the preparation of mass spectrometry samples. Therefore, great care should be taken at every step. The user should be familiar with the Material Safety Data Sheet (MSDS) for all the chemicals and full PPE (safety goggles, lab coat, and gloves) should be worn while working with chemicals and biological materials. Biological materials should be handled inside biosafety cabinets or in a fume hood. Safety caps should be used for centrifugation steps to prevent spillage. All equipment which gets into contact with biological materials should be disinfected with appropriate agents.

6. Summary

In this chapter, we have provided the commonly used protocols for label-free LC-MS/MS proteomic analysis. We mainly focused on the data acquisition and differential expression analysis of large and complex mass spectrometry data sets using Progenesis QI for Proteomics and Proteome Discoverer. Label-free mass spectrometry—based quantification of proteomes can lead to the discovery of biomarkers for clinical screening as well as the development of novel therapeutic strategies. The label-free quantification followed by functional annotations such as gene ontology analysis can also help researchers to understand disease mechanisms and metabolic processes related to disease formation.

7. Notes

Note 1. Collected cells can be used directly by proceeding with the membrane fractionation, or they can be snap frozen by using liquid nitrogen and kept at −80°C until all samples are ready.

Note 2. Cell wash solution provided with MemPer Plus membrane extraction kit to wash cells instead of PBS.

Note 3. Progenesis QI for Proteomics uses a selected or assessed/selected run to compensate for drifts in retention time and aligns all the runs to this chosen reference run.

Note 4. In Progenesis QI for Proteomics automatic processing provides various options. Alignment reference can be chosen by assessing all runs in the experiment for the suitability; a run which is the most suitable run from some of the manually selected candidates can be used or a specific run can be used. For example, in an experimental design which contains three independent experiments from parental and resistant cells.

Option a. Assess all runs in the experiment for suitability will search all six imported data for their suitability to choose a reference.

Option b. Use the most suitable run from the candidates selected by the user. Candidates such as Parental_A, Parental_B, and Parental_C can be chosen manually. In this case, the software will assess only these three imported runs to select the most suitable reference.

Option c. Use this run. In this option, the user can define which run to be selected as a reference for the experiment, for example, Parental_A.

Note 5. The user can identify the reference themselves and prefer to manually align the data. If the user has prior knowledge of the data and has an idea about which run will work best as a reference, they can pick this run and all the other runs are aligned according to the selected reference. However, low-quality chromatography of the user-selected run can reduce the alignment of the runs, and also if the run does not import correctly into Progenesis QI, the alignment will fail. However, if the chromatography is good and the data gets imported correctly, this is the fastest option for alignment. Assessing all runs is a great option when there is no prior knowledge of the data and the software algorithm will pick the best run for alignment. However, this option takes longer compared to others.

Note 6. Filtering parameters provided here can be adjusted depending on the experiment and the user's choice.

Note 7. At the filtering step, the user can filter the data based on the inside area (m/z from X to Y and minutes from A to B), peptide ion charge (Charge 1, Charge 2, …, Charge N), and the number of isotopes.

Note 8. The user can decide which experimental design to choose based on samples from a given subject appearing in only one condition (between-subject design) or under different conditions (within-subject design). Between-subject design simply groups the samples based on the main factor, for example, parental versus resistant, and the ANOVA calculation assumes that conditions are independent. Within-subject design use comparison between more than one parameter (multiple factors), for example, tumor versus adjacent healthy tissue samples being collected from the patients over the time course.

Acknowledgments

This work was supported by an Irish Research Council Postdoctoral Fellowship (Government of Ireland) award (grant no. GOIPD/2021/463) to Dr. Esen Efeoglu. Figs. 10.1, 10.2 and 10.3 were produced using BioRender.

References

[1] Angel TE, Aryal UK, Hengel SM, Baker ES, Kelly RT, Robinson EW, et al. Mass spectrometry based proteomics: existing capabilities and future directions. Chem Soc Rev May 21, 2012;41(10):3912−28.

[2] Shortreed MR, Wenger CD, Frey BL, Sheynkman GM, Scalf M, Keller MP, et al. Global identification of protein post-translational modifications in a single-pass database search. J Proteome Res November 6, 2015;14(11):4714−20.

[3] Baker ES, Liu T, Petyuk VA, Burnum-Johnson KE, Ibrahim YM, Anderson GA, et al. Mass spectrometry for translational proteomics: progress and clinical implications. Genome Med August 31, 2012;4(8):63.

[4] Schilling B, Rardin MJ, MacLean BX, Zawadzka AM, Frewen BE, Cusack MP, et al. Platform-independent and label-free quantitation of proteomic data using MS1 extracted ion chromatograms in skyline: application to protein acetylation and phosphorylation. Mol Cell Proteomics May 2012;11(5):202−14.

[5] Savage SR, Zhang B. Using phosphoproteomics data to understand cellular signaling: a comprehensive guide to bioinformatics resources. Clin Proteomics July 11, 2020;17(1):27.

[6] Sürmen MG, Sürmen S, Ali A, Musharraf SG, Emekli N. Phosphoproteomic strategies in cancer research: a minireview. Analyst November 9, 2020;145(22):7125−49.

[7] Ramroop JR, Stein MN, Drake JM. Impact of phosphoproteomics in the era of precision medicine for prostate cancer. Front Oncol 2018;8. Available from: https://www.frontiersin.org/articles/10.3389/fonc.2018.00028.

[8] Kaushik P, Curell RVB, Henry M, Barron N, Meleady P. LC-MS/MS-based quantitative proteomic and phosphoproteomic analysis of CHO-K1 cells adapted to growth in glutamine-free media. Biotechnol Lett 2020;42(12):2523−36.

[9] Dunphy K, Dowling P, Miettinen JJ, Heckman CA, Meleady P, Henry M, et al. Phosphoproteomic analysis of primary myeloma patient samples identifies distinct phosphorylation signatures correlating with chemo-sensitivity profiles in an ex vivo drug sensitivity testing platform. Blood November 23, 2021;138:2666.

[10] Bryan L, Henry M, Kelly RM, Lloyd M, Frye CC, Osborne MD, et al. Global phosphoproteomic study of high/low specific productivity industrially relevant mAb producing recombinant CHO cell lines. Curr Res Biotechnol January 1, 2021;3:49−56.

[11] Murphy S, Zweyer M, Henry M, Meleady P, Mundegar RR, Swandulla D, et al. Proteomic profiling of liver tissue from the mdx-4cv mouse model of Duchenne muscular dystrophy. Clin Proteomics October 29, 2018;15(1):34.

[12] Dowling P, Pollard D, Larkin A, Henry M, Meleady P, Gately K, et al. Abnormal levels of heterogeneous nuclear ribonucleoprotein A2B1 (hnRNPA2B1) in tumour tissue and blood samples from patients diagnosed with lung cancer. Mol Biosyst February 17, 2015;11(3):743−52.

[13] Dowling P, Moran B, McAuley E, Meleady P, Henry M, Clynes M, et al. Quantitative label-free mass spectrometry analysis of formalin-fixed, paraffin-embedded tissue representing the invasive cutaneous malignant melanoma proteome. Oncol Lett November 1, 2016;12(5):3296−304.

[14] Perkel JM. Single-cell proteomics takes centre stage. Nature September 20, 2021;597(7877):580−2.

[15] Chaerkady R, Thuluvath PJ, Kim MS, Nalli A, Vivekanandan P, Simmers J, et al. 18O labeling for a quantitative proteomic analysis of glycoproteins in hepatocellular carcinoma. Clin Proteomics December 2008;4(3):137−55.

[16] Xu D, Zhu X, Ren J, Huang S, Xiao Z, Jiang H, et al. Quantitative proteomic analysis of cervical cancer based on TMT-labeled quantitative proteomics. J Proteomics February 10, 2022;252:104453.

[17] Gouw JW, Krijgsveld J, Heck AJR. Quantitative proteomics by metabolic labeling of model organisms. Mol Cell Proteomics January 2010;9(1):11−24.

[18] Nakayasu ES, Gritsenko M, Piehowski PD, Gao Y, Orton DJ, Schepmoes AA, et al. Tutorial: best practices and considerations for mass-spectrometry-based protein biomarker discovery and validation. Nat Protoc August 2021;16(8):3737−60.

[19] Dupree EJ, Jayathirtha M, Yorkey H, Mihasan M, Petre BA, Darie CC. A critical review of bottom-up proteomics: the good, the bad, and the future of this field. Proteomes July 6, 2020;8(3):14.

[20] Progenesis QI for proteomics [Internet]. [cited 2022 October 6]. Available from: https://www.nonlinear.com/progenesis/qi-for-proteomics/.

[21] MaxQuant [Internet]. [cited 2022 October 6]. Available from: https://www.maxquant.org/.

[22] PreOmics iST − protein sample preparation for LC-MS [Internet]. [cited 2022 May 23]. Available from: https://www.preomics.com/ist-label-free.

Profiling of the phosphoproteome using tandem mass tag labeling

11

Katie Dunphy[1,2] and Paul Dowling[1,2]

[1]*Department of Biology, Maynooth University, National University of Ireland, Maynooth, Ireland;*
[2]*Kathleen Lonsdale Institute for Human Health Research, Maynooth University, National University of Ireland, Maynooth, Ireland*

1. Key resources table

Reagent or resource	Source	Identifier
Primary patient cells OR cells from cell line	Patient material/Cell line	N/A
Chemicals, peptides, and recombinant proteins		
Sodium dodecyl sulfate (SDS)	Sigma	Cat# L3771
Tris	Sigma	Cat# T6066
Dithiothreitol (DTT)	Sigma	Cat# D9163
Protease inhibitors	Cell Signaling Technology	Cat# 5871
Phosphatase inhibitors	Cell Signaling Technology	Cat# 5870
Hydrochloric acid	Sigma	Cat# 258148
Urea	Sigma	Cat# U0631
Iodoacetamide (IAA)	Sigma	Cat# I1149
HEPES	Sigma	Cat# H3375
Sodium hydroxide (NaOH)	Sigma	Cat# S8045
Anhydrous acetonitrile	Sigma	Cat# 271004
LC-MS grade water	Thermo Fisher Scientific Inc.	Cat# 51140
Trifluoroacetic acid	Sigma	Cat# T6508
Critical Commercial assays		
Pierce 660 nm Protein Assay Kit	Thermo Fisher Scientific Inc.	#22662
TMT10plex Isobaric Mass Tagging Kit	Thermo Fisher Scientific Inc.	#90113
High-Select Fe-NTA Phosphopeptide Enrichment Kit	Thermo Fisher Scientific Inc.	#A32992

Continued

Proteomics Mass Spectrometry Methods. https://doi.org/10.1016/B978-0-323-90395-0.00010-3

Reagent or resource	Source	Identifier
Software and Algorithms		
MaxQuant	https://www.maxquant.org/	N/A
Perseus	https://www.maxquant.org/perseus/	N/A
Other		
Vivacon 500, 30,000 MWCO Hydrosart	Sartorius	Cat# VN01H22
Eppendorf DNA LoBind tubes, 1.5 mL	Sigma	Cat# EP0030108051
Eppendorf DNA LoBind tubes, 2 mL	Sigma	Cat# EP0030108078

2. Materials and equipment

2.1 Cell lysis and protein quantitation

- 1X Lysis Buffer

  4% SDS
  100 mM Tris
  0.1 M Dithiothreitol (DTT)
  Protease inhibitors
  Phosphatase inhibitors
  Bring to 100 mL with dH_2O, stir to dissolve and adjust pH to 7.6 with HCl.
  (Note: Add fresh protease and phosphatase inhibitors to the appropriate amount of lysis buffer required before cells are lysed.)
  Store at 4°C.
- Pierce 660 nm Protein Assay Kit (Thermo Fisher Scientific Inc.).

2.2 Filter-aided sample preparation

- Tris/HCl (0.1 M, pH 8.5)

  1.21g Tris
  Bring to 100 mL with dH_2O, stir to dissolve and adjust pH to 8.5 with HCl.
- Urea (8 M)
  24.02g Urea
  Bring to 50 mL with 0.1 m Tris/HCl pH 8.5.
- Iodoacetamide (0.05 M)
  0.0278g Iodoacetamide
  Bring to 3 mL with 8 m urea.
- HEPES (0.05 M, pH 8.5)
  0.596g HEPES
  Bring to 50 mL dH_2O, stir to dissolve and adjust pH to 8.5 with NaOH, if needed.

2.3 **Tandem mass tag labeling**

- TMT10plex Isobaric Label Reagent Set (Thermo Fisher Scientific Inc.)
- Anhydrous acetonitrile
- LC-MS Grade water
- 100 mM triethyl ammonium bicarbonate (TEAB)
- 5% Hydroxylamine

2.4 **C18 sample clean up**

- Pierce C18 Spin Column (Thermo Fisher Scientific Inc.)
- Activation solution:
   50% acetonitrile (ACN)
- Equilibration and Wash solution:
   0.25% Trifluoroacetic acid (TFA)
   2.5% ACN
- Elution buffer:
   80% ACN

2.5 **Phosphopeptide enrichment**

- High-Select Fe-NTA Phosphopeptide Enrichment Kit (Thermo Fisher Scientific)

2.6 **Equipment**

- Centrifuge (Eppendorf 5417R Refrigerated Centrifuge)
- Thermomixer (Eppendorf Thermomixer Comfort)
- High-resolution mass spectrometer (Thermo Orbitrap Fusion Tribrid Mass Spectrometer (Thermo Scientific))
- UltiMate 3000 nanoRSLC system (Thermo Scientific)
- Speed vacuum (Genevac miVac Centrifugal Concentrators)

---

3. **Experimental procedure**
3.1 **Cell lysis**

Timing: 1–2 h.

1. Resuspend cell pellets in lysis buffer (approximately 100 μL per 1,000,000 cells).
2. Incubate at 95°C for 3 min.
3. Sonicate samples for 10 s then place on ice for 30 s.
4. Repeat this process for a total of three sonications.
5. Incubate samples for 5 min on ice.
6. Centrifuge at 10,000 × g for 20 min to clarify the lysate.
7. Transfer the supernatant to a new tube.

   Pause point: Cell lysates can be stored at −80°C.

3.2 **Protein quantitation**

Timing: 1 h

1. The protein concentration of each sample is determined using the Pierce 660 nm Protein Assay Kit (Thermo Fisher Scientific).
2. Add one pack of Ionic Detergent Compatibility Reagent (IDCR) to 20 mL of Pierce 660 nm Protein Assay Reagent.
   Note 1: The lysis buffer used in this protocol contains SDS, an interfering substance of the Pierce 660 nm Protein Assay. The IDCR must be used to ensure there is no interference during protein quantitation. For samples which do not contain an interfering substance, the IDCR is not required or another protein quantitation assay, such as the Bradford assay, may be used.
3. Add 10 μL of each of seven BSA standards (2000 μg, 1500 μg, 1000 μg, 750 μg, 500 μg, 250 μg, 125 μg) and a blank (deionized H_2O) to a flat-bottomed 96 well plate in triplicate.
4. Add 10 μL of each sample to the plate in triplicate.
5. Add 150 μL of the IDCR solution to each well.
6. Cover the plate and mix on a plate shaker at 300 rpm for 1 min, followed by incubation at room temperature for 5 min.
7. Read absorbance using a spectrophotometer at 660 nm.
8. Create a standard curve and determine the protein concentration of each sample using the preferred method such as the online tool, MyAssays Online.
   Pause point: Cell lysates can be stored at $-80°C$.

3.3 **Filter-aided sample preparation**

Timing: 2 days

1. Place filter unit in the collection tube (Vivacon 500, 30,000 MWCO Hydrosart, Sartorius).
2. Add 200 μL of 8 M Urea to each filter unit.
3. Mix up to 30 μL of protein extract with urea in each filter unit and centrifuge at $14,000 \times g$ for 15 min.
   Note 2: The amount of protein added is subject to the capacity of the filter unit. Ensure samples are sufficiently concentrated to load the desired amount of protein for the analysis. 25 μg of protein from each sample was loaded for our analysis.
4. Add 200 μL of urea to the filter units and centrifuge at $14,000 \times g$ for 15 min.
5. Discard the flowthrough.
6. Add 100 μL of 50 mM iodoacetamide (IAA) to the filter unit.
7. Mix at 600 rpm for 1 min in a thermo-mixer and incubate at room temperature for 20 min. Cover samples during incubation as IAA is a light-sensitive compound.
8. Centrifuge samples at $14,000 \times g$ for 10 min.

9. Add 100 μL of urea and centrifuge at 14,000 × g for 15 min. Repeat this step twice.
10. Add 100 μL of 50 mM HEPES, pH 8.5 to the filter units and centrifuge at 14,000 × g for 10 min. Repeat this step twice.
11. Add 40 μL of HEPES mixed with trypsin (1:25 enzyme-to-protein ratio) to the filter units and mix at 600 rpm in a thermomixer for 1 min.
12. Incubate samples in a wet chamber at 37°C for 4–18 h.
    Note 3: To prepare the wet chamber, insert tissue (four to five layers) into a box and wet the tissue with water. Insert a microcentrifuge rack containing the samples. Place the lid on the box and incubate at 37°C.
13. Transfer the filter units to new collection tubes and centrifuge at 14,000 × g for 10 min.
14. Add 40 μL of HEPES to the filter units and centrifuge at 14,000 × g for 10 min.
    Note 4: Do not acidify samples at this point, continue to the TMT labeling protocol.

3.4 Tandem mass tag labeling

Timing: 3–6 h.

1. Equilibrate both sets of TMT label reagents to room temperature.
2. Add 41 μL of anhydrous acetonitrile to each reagent tube. Allow tube to stand for 5 min with occasional vortexing to ensure reagents are dissolved.
3. Add 10 μL of the appropriate TMT label reagent to each protein digest.
    Note 5: The amount of TMT label reagent added to each protein digest is dependent on the protein concentration of the sample. Here, a TMT label reagent to protein ratio of ∼8:1 was used. Zecha et al. provide more detailed information on the optimization of TMT label reagent-to-peptide ratios [2].
4. Incubate the reaction at room temperature for 1 h with occasional vortexing.
    Note 6: Ideally, a "label check" is performed prior to quenching the reaction. This ensures that there is a high labeling efficiency and equal amounts of protein present in each sample [3].
5. Add 2 μL of 5% hydroxylamine to each sample and incubate for 15 min to quench the reaction.
6. Combine 90 μL from each sample of both TMT plexes to obtain two sample preparations containing 10 peptide labeled samples in each.
7. Completely or partially dry the pooled samples. If partially dried, acidify the remaining sample in a 1:7 ratio (1-part acidic sample buffer, 7 parts sample) using 2% TFA, 20% ACN. If completely dried, resuspend sample in 0.25% TFA, 2.5% ACN.
    Pause Point: Pooled samples can be stored at 4°C overnight.

3.5 **C18 sample clean-up**

Timing: 4–5 h.

1. Determine the number of spin columns required based on the binding capabilities of the columns.
   Note 7: Each Pierce C18 spin column can bind up to 30 μg of total peptide. Depending on the predicted concentration of your sample, use the sufficient number of columns required.
2. Place columns in a receiver tube and add 200 μL of Activation Buffer (50% ACN).
3. Centrifuge at 1500 × g for 1 min.
4. Discard flowthrough and repeat this step once.
5. Add 200 μL of Equilibration Solution (0.25% TFA in 2.5% ACN) to the column and centrifuge at 1500 × g for 1 min.
6. Discard the flowthrough and repeat this step once.
7. Divide the amount of sample equally between the sufficient number of C18 spin columns based on the sample concentration.
8. Add columns to fresh receiver tubes and centrifuge at 1500 × g for 1 min.
9. Recover the flowthrough and repeat this step once.
10. Place column in a receiver tube. Add 200 μL of Wash Solution (0.25% TFA in 2.5% ACN) to the column.
11. Centrifuge at 1500 × g for 1 min.
12. Discard the flowthrough and repeat this step once.
13. Place the column in a fresh tube. Add 20 μL of Elution Buffer (80% ACN) to the column and centrifuge at 1500 × g for 1 min.
14. Centrifuge at 1500 × g for 1 min and repeat this step once.
15. Dry samples completely in a speed vacuum concentrator.
    Pause Point: Samples may be stored at −80°C at this point.

3.6 **Phosphopeptide enrichment**

Timing: 5–6 h.

1. Isolate phosphopeptides using the High-Select Fe-NTA Phosphopeptide Enrichment Kit.
2. Resuspend dried peptide eluates in 200 μL of Binding/Wash Buffer.
3. Loosen the screw caps and remove the bottom closures of the columns.
4. Place columns in 2 mL microcentrifuge collection tubes and centrifuge at 1000 × g for 30 s to remove storage buffer.
5. Remove the screw caps. Add 200 μL of Binding/Wash Buffer to each column and spin at 1000 × g for 30 s.
6. Discard the flowthrough and repeat this step once.
7. Place the white Luer plugs on the bottom of the columns and place in empty 2 mL microcentrifuge tubes.

8. Add 200 μL of each suspended peptide sample to each equilibrated column.
9. Close screw caps and mix the resin with the sample by carefully tapping the bottom plug while holding the screw cap until the resin is in suspension (15−20 s).
10. Incubate the samples for 30 min with gentle mixing every 10 min.
11. Place the column in a microcentrifuge tube, spin at $1000 \times g$ for 30 s and **recover** the flowthrough.
12. Add 200 μL of Binding/Wash Buffer and centrifuge at $1000 \times g$ for 30 s.
13. Repeat this step for a total of three washes and **recover** the flowthrough.
14. Add 200 μL of LC-MS/MS grade water, centrifuge at $1000 \times g$ for 30 s and **discard** the flowthrough.
15. Place the columns in new microcentrifuge tubes.
16. To elute the phosphopeptides, add 100 μL of Elution Buffer to each column and centrifuge at $1000 \times g$ for 30 s.
17. Repeat this step once.
18. Dry the eluate immediately in a speed vacuum.
19. Resuspend the dried eluate in 0.25% TFA, 2.5% ACN.
20. Combine the recovered flowthrough from steps 11 and 13 to make up the un-enriched protein fraction.
21. Acidify the unphosphorylated fractions with one part acidification buffer (2% TFA, 20% ACN), seven parts sample.

Pause Point: Samples may be stored at −20 or −80°C at this point.

3.7 Mass spectrometry

1. Resuspended labeled peptides are picked up by the autosampler and loaded onto the trapping column (PepMap100, C18, 300 μm × 5 mm) (Thermo Scientific) for 3 min at a flow rate of 25 μL/min. Peptides are then resolved on an analytical column (Easy-Spray C18 75 μm × 250 mm, 2 μm bead diameter column) (Thermo Scientific) using a gradient of 98% A (0.1% [v/v] formic acid (FA)): 2% B (80% [v/v] ACN, 0.08% [v/v] FA) to 35% B over 120 min at a flow rate of 300 μL/min.
2. Set the Orbitrap Fusion Tribrid to acquire the MS1 spectra over m/z 200−1500 in the Orbitrap (120 K resolution at 200 m/z), and set the automatic gain control (AGC) to accumulate 4×10^5 ions with a maximum injection time of 50 ms. Data-dependent tandem MS analysis can be carried out using a top-speed approach (cycle time of 3 s). MS2 spectra are acquired in the ion trap. The intensity threshold for fragmentation is set to 5000 and includes charge states 2+ to 6+.
3. Phosphosite identification is achieved by enabling multistage activation (MSA) for neutral-loss triggered fragmentation for all precursor ions exhibiting neutral loss of mass 97.9763 with a mass tolerance of 0.5 m/z, where the neutral loss ion is one of the top 10 most intense MS2 ions. Set the AGC to 20,000 with a maximum injection time set at 90 ms.

4. Expected outcomes

This protocol allows users to isolate, identify, and quantify proteins and phosphoproteins from primary animal cells or cell lines in a robust manner. The protein yield following cell lysis will vary depending on the number of cells lysed and cell type. This must be considered prior to starting this protocol. The TMT labeling efficiency and resolution of the mass spectrometer will also affect the number of protein identifications. In our case, 25 μg of protein from each sample was subject to trypsin digestion and further analysis. Our phosphoproteomic analysis identified a total of 2945 phosphorylation sites and our proteomic analysis identified a total of 1473 proteins. An increase in the amount of starting protein material would presumably result in an increased number of identifications.

5. Quantification and statistical analysis

The identification and quantitation of proteins in the unenriched fraction and phosphoproteins in the phospho-enriched fraction can be performed using a variety of software including MaxQuant, Proteome Discoverer, and Progenesis QI for Proteomics (Waters). MaxQuant (www.maxquant.org) is a powerful, freely available software with a built-in search engine, Andromeda, used for the analysis of high-resolution mass spectrometry data. A database search can be performed in MaxQuant using a human reference proteome FASTA database (available online at www.uniprot.org). In MaxQuant, peptide sequences identified from the mass spectra are assembled into a list of proteins. Raw files are uploaded to MaxQuant and unique values are entered for each sample in the Experiment column. For the phospho-enriched fractions, set the PTM parameter to "True." In the Group-Specific Parameters pane, change Type to "Reporter ion MS3" and select "TMT 10-plex." In the Modifications tab, set "Phospho (STY)" as a variable modification. In the Global Parameters pane, upload the chosen FASTA database. After the MaxQuant run is complete, ".txt" files are generated which contain the TMT reporter ion intensities as well as additional information providing insight into the confidence of the peptide and protein identifications. For data wrangling, normalization, and statistical analysis, ".txt" files can be easily uploaded to the Perseus framework (https://www.maxquant.org/perseus/). Normalization of TMT reporter ion intensities can be performed by median subtraction. If the study design includes common pooled standards for each TMT plex, a more robust method of normalization is the use of the Internal Reference Scaling (IRS) method as previously described by Gupta et al. and Plubell et al. [4,5]. Following normalization, statistical analysis can be performed in Perseus to identify differentially abundant proteins and phosphorylation sites between the conditions being analyzed [6]. Tutorials can be found on the Perseus website (https://www.maxquant.org/perseus/). Perseus allows users to annotate proteins, perform statistical tests and create informative scientific figures such as histograms, volcano plots, and heatmaps.

6. Advantages

TMT labeling is an important approach for quantitative proteomics. A significant advantage in this approach is the ability to employ a multiplexing strategy, that allows for higher throughput for quantitative analyses with meaningful proteome coverage. TMT labeled samples can be combined into a single sample for subsequent separation, mass spectrometry, and data analysis.

7. Limitations

Disadvantages of the described labeling method includes the possibility of inconsistency in labeling efficiencies and price of the reagents. The protocol for sample preparation involves multiple steps to achieve reproducible and consistent results. With increasing multiplexicity of labeling reagents, such as TMT, the need for state-of-the-art high resolution mass spectrometer is unavoidable to produce high quality data.

8. Optimization and troubleshooting

TMT labeling efficiency should be examined to ensure all samples are fully labeled by TMT before proceeding with the main experiment of labeling all samples and subsequent mass spectrometry analysis.

A premix ratio test should be performed to confirm that the correct ratio for equal mixing is correct before pooling together the experimental samples prior to analysis. Mix a small aliquot of four to five TMT-labeled sample together and analyze by LC-MS/MS. Repeat this experiment, keeping one to two samples the samples, but introduce new samples to the mix. Use this approach to confirm the consistency of detection and quantitation.

9. Safety considerations and standards

Wear appropriate protective eyewear, clothing, and gloves. To avoid contamination of samples for mass spectrometry analysis, always wear gloves when handling samples and gels. Use ultrapure mass spectrometry grade reagents. Perform sample preparation in a clean work area.

10. Alternative methods/procedures

Alternative approaches for labeling include use of isobaric tags for relative and absolute quantitation (iTRAQ) reagents. iTRAQ technology utilizes isobaric reagents to label the primary amines of peptides and proteins. The iTRAQ reagents usually consist of an N-methyl piperazine reporter group, a balance group, and an N-hydroxy succinimide ester group that is reactive with the primary amines of peptides.

An alternative for phosphopeptide enrichment would be by immunoprecipitation using a specific bead-conjugated antibody in conjunction with liquid chromatography tandem mass spectrometry (MS/MS) for quantitative profiling of PTM sites. This approach allows for the selective enrichment of phospho-tyrosine and/or phospho-serine/threonine-modified peptides.

References

[1] Brenes A, Hukelmann J, Bensaddek D, Lamond AI. Multibatch TMT reveals false positives, batch effects and missing values. in eng Mol Cell Proteomics October 2019; 18(10):1967—80. https://doi.org/10.1074/mcp.RA119.001472.

[2] Zecha J, Satpathy S, Kanashova T, Avanessian SC, Kane MH, et al. TMT labeling for the masses: a robust and cost-efficient, in-solution labeling approach. in eng Mol Cell Proteomics July 2019;18(7):1468—78. https://doi.org/10.1074/mcp.TIR119.001385.

[3] Navarrete-Perea J, Yu Q, Gygi SP, Paulo JA. Streamlined tandem mass tag (SL-TMT) protocol: an efficient strategy for quantitative (Phospho)proteome profiling using tandem mass tag-synchronous precursor selection-MS3. in eng J Proteome Res June 1, 2018; 17(6):2226—36. https://doi.org/10.1021/acs.jproteome.8b00217.

[4] Gupta R, Min CW, Kim YJ, Kim ST. Identification of Msp1-induced signaling components in rice leaves by integrated proteomic and phosphoproteomic analysis. in eng Int J Mol Sci August 24, 2019;20(17). https://doi.org/10.3390/ijms20174135.

[5] Plubell DL, Wilmarth PA, Zhao Y, Fenton AM, Minnier J, et al. Extended multiplexing of tandem mass tags (TMT) labeling reveals age and high fat diet specific proteome changes in mouse epididymal adipose tissue. in eng Mol Cell Proteomics May 2017;16(5): 873—90. https://doi.org/10.1074/mcp.M116.065524.

[6] Tyanova S, Temu T, Sinitcyn P, Carlson A, Hein MY, et al. The perseus computational platform for comprehensive analysis of (prote)omics data. Nat Methods 2016/09/01 2016;13(9):731—40. https://doi.org/10.1038/nmeth.3901.

Phosphopeptide enrichment and post translational modification characterization using LC-MS/MS

Michael Henry[1] and Paula Meleady[1,2]

[1]*National Institute for Cellular Biotechnology, Dublin City University, Glasnevin, Dublin, Ireland;*
[2]*School of Biotechnology, Dublin City University, Glasnevin, Dublin, Ireland*

1. Introduction

In this chapter, we are outlining protocols for the enrichment of phosphopeptides from Chinese hamster ovary cells followed by differential analysis using both labeled and label-free phosphoproteomic LC-MS/MS approaches. These protocols can be then applied to any mammalian cell line.

Protein modification by phosphorylation can have a profound impact on the activity, conformation, and localization of a phosphoprotein [1]. It plays a critical role in nearly all biological processes. Three in every four proteins are suspected to be phosphorylated in their life cycle: however, phosphorylation stoichiometry at any given time is generally very low [2]. As a result of low stoichiometry, the abundance of phosphorylation is a number of orders of magnitude below that of the unphosphorylated peptide and therefore some form of enrichment is needed to increase phosphopeptide coverage.

Quantitative phosphoproteomic analysis can be achieved with label and label-free approaches. It is critical to purify phosphopeptides prior to analysis by MS. A minimum of half a milligram (0.5 mg) of peptide is recommended for a label-free approach, while pooling of samples using a labeled approach, for example using the TMT10plex Isobaric Mass Tags (Thermo Fisher Scientific), can allow for equivalent peptide concentrations.

Chinese hamster ovary (CHO) cells are the principal host cell line used to produce biopharmaceuticals. Up to 85% of new monoclonal antibody (Mab) products approved between 2014 and 2018 were produced in CHO cells [3]. In recent years, CHO cell research has mainly focused on understanding the genomic, transcriptomic, and to some extent, proteomic changes in these cells under different industrially relevant contexts [4]. A number of recent studies by our group and others have shown the additional benefit of including phosphoproteomic data

Proteomics Mass Spectrometry Methods. https://doi.org/10.1016/B978-0-323-90395-0.00001-2

in CHO cell studies to understand bioprocess relevant phenotypes such as growth and productivity [5–7].

2. Materials and equipment

- Mass Spectrometry Grade Lys-C Protease (Fisher Scientific 15561995)
- Mass Spectrometry Grade Trypsin (Fisher Scientific 15255753)
- Halt Protease Inhibitor Cocktail (100X) (Fisher Scientific 78430)
- Halt Phosphatase Inhibitor Cocktail (Fisher Scientific 78427)
- Urea ACS reagent 99%–100% (Sigma–Aldrich U5128)
- Ammonium Bicarbonate BioUltra, ≥99.5% (Sigma–Aldrich 09830)
- Trizma acetate, ≥99.5% (Sigma–Aldrich T1258)
- Trizma hydrochloride, ≥99.5% (Sigma–Aldrich T3253)
- Sodium chloride, ≥99.5% (Sigma–Aldrich S9888)
- Phosphate Buffered Saline (Sigma–Aldrich D8537)
- DL-Dithiothreitol (DTT) BioUltra (Sigma–Aldrich 43815)
- Iodoacetamide BioUltra (Sigma–Aldrich I1149)
- Trifluoroacetic Acid (TFA), Optima LC/MS Grade, Fisher Chemical (Fisher Scientific 10723857)
- Formic Acid (FA), 99.0+%, Optima LC/MS Grade, Fisher Chemical (Fisher Scientific 10780320)
- Water, Optima LC/MS Grade, Fisher Chemical (Fisher Scientific 10728098)
- Acetonitrile (ACN), Optima LC/MS Grade, Fisher Chemical (Fisher Scientific 10489553)
- Liquid Nitrogen/Dry ice for snap freezing
- Thermo Scientific Pierce Fe-NTA Phosphopeptide Enrichment Kit (Fisher Scientific 10389287)
- Thermo Scientific TMT10plex Isobaric Label Reagent Set, 1 × 0.8 mg (Fisher Scientific 90110)
- Thermo Scientific Pierce BCA Protein Assay Kit (Fisher Scientific 23225)
- PreOmics iST 8× (PreOmics 00001)
- Universal pH indicator strips (VWR 1.09535.001)
- Air-circulation thermostat
- Bench-top centrifuge
- Vacuum centrifuge
- Fumehood
- Micropipettes
- Top pan weighing balance
- Sonicating water bath
- Sonicating probe
- Low Protein Binding Microcentrifuge Tubes (Fisher Scientific 90410)
- SpeedVac Vacuum dryer
- Freeze Dryer

- A nanoflow ultra-high performance liquid chromatography (UHPLC) instrument: e.g., UltiMate 3000 RSLCnano System (Thermo Scientific)
- High resolution, high sensitivity, fast Scanning MS instrument: e.g., Orbitrap Fusion Tribrid Mass Spectrometer (Thermo Scientific)
- Peptide trapping Column: Acclaim PepMap100, C18, 300 μm × 5 mm (Fisher Scientific 160454)
- Peptide resolving Column: Acclaim PepMap 100, 75 μm × 50 cm, 3 μm bead diameter column (Fisher Scientific 164570)
- Nanodrop One spectrophotometer (Labtech International)
- MS database search program and phosphorylation confidence site localization software, e.g., SEQUEST HT and ptmRS [8]
- Protein sequence database, e.g., NCBI protein FASTA sequence database
- Quantitative software analysis, e.g., Progenesis QI for Proteomics (Waters), Proteome Discoverer (Thermo Scientific)

3. Step-by-step method details

3.1 Protein preparation and in-solution protein digestion for label-free phosphoproteomic analysis

The methodology is described in detail in this section and a schematic of the work for the preparation of a cell pellet using an in-solution digestion prior to phospho-peptide enrichment and label-free analysis is provided in Fig. 12.1. As an example of an experimental workflow, six CHO-K1 cell line samples in total will be prepared (3 control and 3 test samples).

1. Prepare the six CHO-K1 cell pellets at a cell density of $2x10^7$ cells by centrifuging at 3000 ×g for 5 min and remove the supernatant.
2. Wash the pellets with ice cold PBS and centrifuge at 3000 ×g for 5 min. Discard wash and repeat this step.
3. Resuspend each cell pellet in 1 mL of 8 M urea, 50 mM Tris, 75 mM NaCl (pH 8.2) buffer containing 1× Halt Protease Inhibitor Cocktail and 0.01% ProteaseMAX surfactant detergent.
4. Disrupt each cell pellet by sonication using a micro tip probe ultrasonicator. Place the tube on ice. Set the sonicator to a medium setting and use 30 s sonication pulses allowing to briefly cool in between pulses for a total of three times (Note 1).
5. Spin the lysates at 14,000 ×g for 10 min using a refrigerated centrifuge and transfer the supernatant into a clean microcentrifuge tube.
6. Use the BCA assay to determine protein concentration as per manufacturer's instructions and aliquot protein samples as 1000 μg lots. Aliquoted protein samples can be prepared for digestion straight away or kept at −80°C until required.

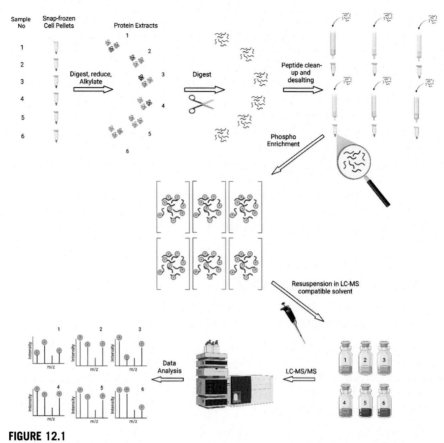

FIGURE 12.1

General workflow for the label-free LC-MS/MS proteomic analysis of phosphopeptide-enriched samples.

Created with BioRender.com.

7. Prepare a 0.5 M dithiothreitol (DTT) stock solution and add to a final concentration of 5 mM DTT for each 1 mg of lysate sample and incubate at 56°C for 20 min (Note 2).

8. Allow the samples to cool down to room temperature.

9. Add freshly prepared iodoacetamide to a final concentration of 15 mM to each sample and incubate at room temperature for 15 min in the dark (Note 3).

10. Adjust the concentration of urea in the sample to less than 1.2 M urea using 25 mM Tris—HCl (pH 8.2) prior to the digestion (Note 4).

11. Add Lys-C, mass spectrometry grade, for an enzyme: protein ratio of 1:200 and incubate at 37°C for 6—18 h (Note 5).

12. Add Trypsin, mass spectrometry grade, at the desired enzyme: substrate ratio of 1:100 and incubate at 37°C for 2—18 h.

13. After digestion, acidify the peptides sample with the addition of TFA to a final concentration of 0.1%.
14. Use a Sep-Pak C18 Classic Cartridge, 360 mg Sorbent per Cartridge, 55–105 μm to concentrate and desalt the peptide sample.
15. Add 9 mL of ACN to condition the C18 cartridge and discard the flowthrough.
16. Add 9 mL of 0.1% TFA to equilibrate the column and discard the flowthrough.
17. Add the entire peptide sample to the column and reapply the flowthrough once (Note 6).
18. Wash the cartridge with 9 mL of 0.1% TFA.
19. Elute the peptides from the cartridge with 5 mL of 50% ACN containing 0.1% TFA.
20. Repeat for the remaining five samples.
21. Snap freeze (flash freeze) the eluted peptides using dry ice or liquid nitrogen and transfer to a lyophilizer to evaporate until dryness.
22. Once dried, the samples can be stored at −80°C or prepared for phosphopeptide enrichment.

3.2 Protein preparation and in-solution protein digestion for labeled phosphoproteomic analysis using the PreOmics iST kit

The methodology is described in detail in this section and a schematic of the workflow for the preparation of small cell pellets for in-solution digestion using the PreOmics iST kit prior to TMT labeling and phosphopeptide enrichment is provided in Fig. 12.2. Six samples in total will be prepared (3 control and 3 test samples).

1. Prepare six CHO-K1 cell pellets at a cell density of 6×10^5 cells by centrifuging at 3000 ×g for 5 min and remove the supernatant.
2. Add 50 μL of lysis buffer to each cell pellet and place the tube in a heating block at 95°C; 1000 rpm; 10 min and then allow to cool to room temperature.
3. Transfer the solubilized protein to the cartridge and add 50 μL of digest buffer. Place the cartridge/tube in a heating block at 37°C; 500 rpm; 1–3 h.
4. Stop the enzymatic reaction with 100 μL stop solution. At room temperature; 500 rpm; 1 min.
5. Spin the cartridge using a centrifuge at 3800 ×g for 3 min.
6. Wash peptides with 200 μL of wash 1 solution and spin the cartridge using a centrifuge at 3800 ×g for 3 min.
7. Wash peptides with 200 μL of wash 2 solution and spin the cartridge using a centrifuge at 3800 ×g for 3 min.
8. Transfer the cartridge to a new tube and add 100 μL of elute buffer. Spin the cartridge using a centrifuge at 3800 ×g for 3 min. Repeat once keeping the flowthrough peptides.
9. Transfer the tube containing pooled flowthrough peptides to a SpeedVac dryer and centrifuge at 48°C until complete dryness. The dried sample can be stored at −80°C or prepared for TMT labeling.

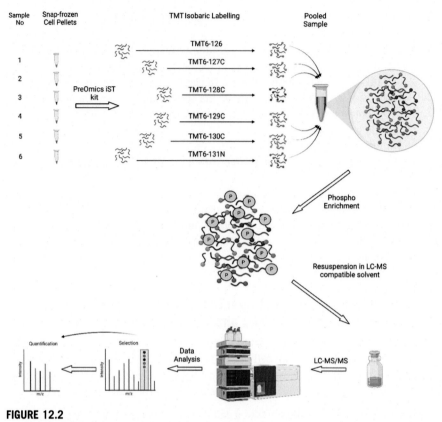

FIGURE 12.2

General workflow for TMT labeled peptides followed by phosphopeptide enrichment and LC-MS/MS proteomic analysis.

Created with BioRender.com.

3.3 TMT labeling

The methodology is described in detail in this section and a schematic of the workflow for TMT labeling six samples digested following in-solution digestion using the PreOmics iST kit is shown in Fig. 12.2. From a TMT10plex Isobaric Label Reagent set we used TMT^{10}-126, 127C, 128C, 129C, 130C, and 131N −0.8 mg label vials (six in total).

1. Immediately before use, equilibrate the TMT Label Reagents to room temperature.
2. If the peptide samples were stored at −80°C, also equilibrate to room temperature.
3. Add 41 μL of anhydrous acetonitrile to each TMT vial and allow the reagent to dissolve for 5 min with occasional vortexing. Briefly centrifuge each tube to gather the solution.

4. Add 41 μL of the TMT Label Reagent to each dried peptide sample.
5. Incubate the reaction of peptide and TMT reagent for 1 h at room temperature.
6. Add 8 μL of 5% hydroxylamine to each sample and incubate for 15 min to quench the reaction.
7. Combine the six samples in a new microcentrifuge tube and store at −80°C.
8. Use a Sep-Pak C18 Classic Cartridge, 360 mg Sorbent per Cartridge, 55−105 μm to concentrate and desalt the TMT-labeled peptide sample.
9. Add 9 mL of ACN to condition the C18 cartridge and discard the flowthrough.
10. Add 9 mL of 0.1% TFA to equilibrate the column and discard the flowthrough.
11. Add the TMT-labeled sample to the column and reapply the flowthrough once (Note 6).
12. Wash the cartridge with 9 mL of 0.1% TFA.
13. Elute the peptides from the cartridge with 5 mL of 50% ACN containing 0.1% TFA.
14. Repeat for the remaining five samples.
15. Snap freeze (flash freeze) the eluted peptides using dry ice or liquid nitrogen and transfer to a lyophilizer to evaporate until dryness.
16. Once dried, the sample can be stored at −80°C or prepared for phosphopeptide enrichment.

3.4 **Phosphopeptide enrichment**

The methodology is described in detail in this section is for phosphopeptide enrichment from label-free peptide preparation or from TMT labeled peptide preparation.

1. If the dried peptide samples were stored at −80°C, equilibrate to room temperature.
2. Resuspend the peptide sample in 200 μL of the Binding buffer and vortex to completely dissolve the sample.
3. Check the pH using pH paper as the optimal pH for binding is <3 (adjust if necessary).
4. Prepare the phosphopeptide spin tube by removing the bottom closure and slightly opening the screw cap.
5. Transfer the tube to a 2 mL microcentrifuge tube and centrifuge at 1000 ×g for 30 s to remove the storage buffer.
6. Add 200 μL of wash buffer and centrifuge at 1000 ×g for 30 s. Discard the flowthrough and repeat the step.
7. Replace the bottom closure on the spin tube and add the 200 μL of peptide sample. Gently mix the peptide sample and resin by tapping the tube.
8. Incubate for 30 min at room temperature and gently tap the mixture every 10 min to mix the suspension.
9. Transfer the spin tube to a new collection tube and centrifuge at 1000 ×g for 30 s.

10. Transfer the spin tube to a new wash collection tube, add 200 μL of wash buffer and centrifuge at 1000 ×g for 30 s. Discard the flowthrough and repeat the step for two additional washes.
11. Wash the spin tube with 200 μL of LC-MS water and centrifuge at 1000 ×g for 30 s.
12. Elute the phosphopeptides from the spin column with 100 μL of Elution buffer and centrifuge at 1000 ×g for 30 s. Repeat this step once.
13. Immediately dry the eluted phosphopeptide sample (Note 7).
14. The dried phosphopeptides can be stored at −80°C, and resuspended in an LC-MS suitable solution for LC-MS/MS analysis.
15. Following resuspension, the phospho-enriched peptides can be quantified using a Nanodrop One spectrophotometer.

3.5 LC-MS analysis

Use an UltiMate 3000 nano RSLC (Thermo Scientific) to perform nano-flow reverse-phased capillary high pressure liquid chromatography (HPLC) system in combination with an Orbitrap Fusion Tribrid Mass Spectrometer (MS).

1. Load a 1 μL phosphopeptide sample onto the trapping column (PepMap100, C18, 5 mm) at a flow rate of 25 μL/min with 2% (v/v) ACN, 0.1% (v/v) TFA for 3 min.
2. The sample is then resolved onto an analytical column (Acclaim PepMap100, 75 μm × 50 cm) where a binary gradient is employed to elute the peptides over 120 min at 300 nL/min (LC running conditions are provided in Fig. 12.3).
3. Eluting peptides from the HPLC can be analyzed on a Thermo Scientific Orbitrap Fusion Mass Spectrometry (MS) at top speed.

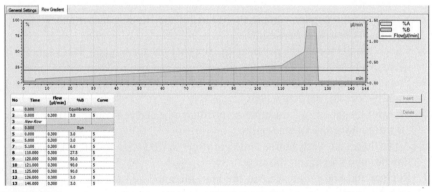

FIGURE 12.3

LC conditions for peptide separation. Buffer A consists of 100% water with 0.1% (v/v) FA and Buffer B consists of 80% ACN with 0.08% (v/v) FA.

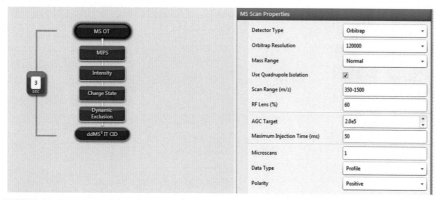

FIGURE 12.4

MS1 conditions for both label and label-free analysis.

4. For MS1 conditions, a voltage of 1.9 kV is applied to a coated capillary emitter tip (Note 8). Full scans within a 380–1500 m/z range are carried out in the Orbitrap mass analyzer at a resolution of 120,000 (at m/z 200), automatic gain control (AGC) target value of 2×10^5 and a maximum ion injection time of 50 ms (MS conditions is provided in Fig. 12.4).

5. MS2 conditions for label-free phosphopeptide analysis uses low resolution scanning multistage activation (MSA) fragmentation technique (which is a pseudo MS3 experiment). In MSA, the neutral loss ion is collisionally activated while the fragments from the precursor ion are still present in the trap which will produce a spectrum that is a combination of MS/MS and MS3 fragmentation (MS conditions are provided in Fig. 12.5).

6. MS2 conditions for TMT-labeled phosphopeptides uses higher-energy-collisional dissociation (HCD) with high resolution 50K scanning with a maximum IT of 90 ms. Phosphopeptide sequence information and reporter ion intensities are present in the one high resolution MS/MS scan. (MS conditions are provided in Fig. 12.6.)

7. An alternative MS method for TMT phosphopeptide analysis is synchronous precursor selection (SPS) mass spectrometry to enable more accurate multiplexed quantification through comparison of reporter ion intensities at the MS3 level. The SPS approach can be implemented on an Orbitrap Tribrid mass spectrometer with selection of the phosphopeptide in the MS scan, followed by MS2 fragmentation by CID in the low-resolution ion trap for phosphopeptide sequence information. Finally, another MS2 fragmentation is carried out with HCD and the resulting MS3 fragments of the reporter ions are then detected in the high resolution Orbitrap analyzer. (MS conditions for SPS are provided in Fig. 12.7.) (Note 9)

FIGURE 12.5

MS2 data-dependent scanning conditions for phosphopeptide label-free analysis using multistage activation fragmentation.

FIGURE 12.6

MS2 data-dependent scanning conditions for TMT-labeled phosphopeptide analysis.

3.6 MS data processing

For label-free phosphopeptide analysis using MSA, Progenesis QI for Proteomics (Waters Company, www.nonlinear.com) and Thermo Scientific Proteome Discoverer 2.2 software using the SEQUEST-HT search engine are used. A precursor mass tolerance of 10 ppm and a fragment mass tolerance of 0.6 Da are set. Carbamidomethylation (+57.021 Da) for cysteine as a fixed modification and with methionine oxidation (+15.996 Da) and phosphorylation (+79.966 Da, T, Y, S) are set. (The search conditions are provided in Fig. 12.8.)

a)

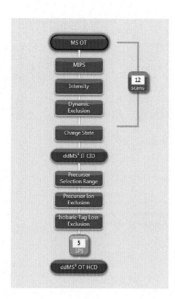

b)

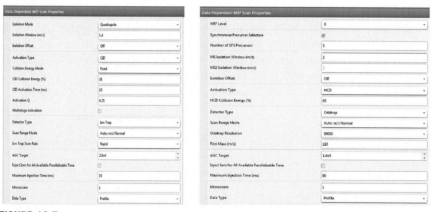

FIGURE 12.7

(A) MS1 conditions for TMT-labeled phosphopeptide analysis using SPS MS3 analysis.
(B) MS2 and MS3 conditions for TMT-labeled phosphopeptide analysis using SPS method.

For TMT phosphopeptide analysis, Thermo Scientific Proteome Discoverer 2.2 software using the SEQUEST HT search engine is used. A precursor mass tolerance of 10 ppm and fragment mass tolerance of 0.02 Da for MS2 conditions (0.6 Da for SPS conditions) is set. Carbamidomethylation (+57.021 Da) of cysteine is set as a fixed modification. Methionine oxidation (+15.996 Da) and phosphorylation

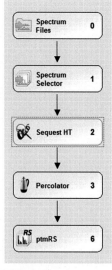

1. Input Data	
Protein Database	Cricetulus_griseus_UniProtKB 2022.fasta
Enzyme Name	Trypsin (Full)
Max. Missed Cleavage Sites	2
Min. Peptide Length	6
Max. Peptide Length	144
2. Tolerances	
Precursor Mass Tolerance	10 ppm
Fragment Mass Tolerance	0.6 Da
Use Average Precursor Mass	False
Use Average Fragment Mass	False
3. Spectrum Matching	
4. Dynamic Modifications	
Max. Equal Modifications Per Peptide	3
1. Dynamic Modification	Oxidation / +15.995 Da (M)
2. Dynamic Modification	Phospho / +79.966 Da (S, T, Y)
3. Dynamic Modification	None
4. Dynamic Modification	None
5. Dynamic Modification	None
6. Dynamic Modification	None
5. Dynamic Modifications (peptide terminus)	
6. Dynamic Modifications (protein terminus)	
7. Static Modifications	
Peptide N-Terminus	None
Peptide C-Terminus	None
1. Static Modification	Carbamidomethyl / +57.021 Da (C)

FIGURE 12.8

Search parameters for MSA analyzed phosphopeptides.

(+79.966 Da, T, Y, S) are set as variable modifications. TMT modification of +229.163 Da is set as a static modification on any N-terminus peptide and lysine peptides. (The search conditions are provided in Fig. 12.9.)

In this case, we search the data against the NCBI *Cricetulus griseus* database protein FASTA database with a 1% FDR criteria using Percolator.

PhosphoRS is used for site localization using >75% probability cut off [8].

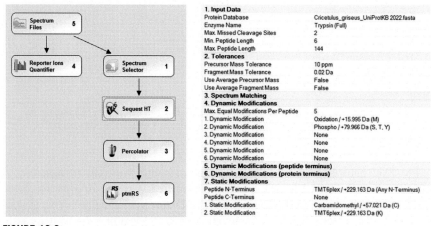

1. Input Data	
Protein Database	Cricetulus_griseus_UniProtKB 2022.fasta
Enzyme Name	Trypsin (Full)
Max. Missed Cleavage Sites	2
Min. Peptide Length	6
Max. Peptide Length	144
2. Tolerances	
Precursor Mass Tolerance	10 ppm
Fragment Mass Tolerance	0.02 Da
Use Average Precursor Mass	False
Use Average Fragment Mass	False
3. Spectrum Matching	
4. Dynamic Modifications	
Max. Equal Modifications Per Peptide	5
1. Dynamic Modification	Oxidation / +15.995 Da (M)
2. Dynamic Modification	Phospho / +79.966 Da (S, T, Y)
3. Dynamic Modification	None
4. Dynamic Modification	None
5. Dynamic Modification	None
6. Dynamic Modification	None
5. Dynamic Modifications (peptide terminus)	
6. Dynamic Modifications (protein terminus)	
7. Static Modifications	
Peptide N-Terminus	TMT6plex / +229.163 Da (Any N-Terminus)
Peptide C-Terminus	None
1. Static Modification	Carbamidomethyl / +57.021 Da (C)
2. Static Modification	TMT6plex / +229.163 Da (K)

FIGURE 12.9

Search parameters for TMT MS2 (Orbitrap analyzer) phosphopeptides.

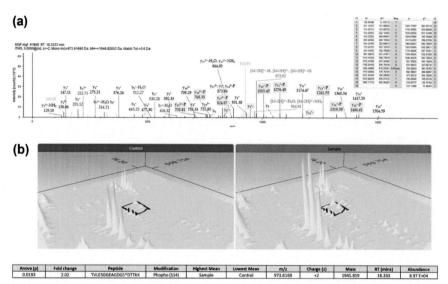

FIGURE 12.10

(A) Phosphopeptide from H/ACA ribonucleoprotein complex subunit 4 confidently identified following phosphopeptide enrichment using MSA MS method and label-free analysis. Sequence: TVLESGGEAGDGDS*DTTKK, S14-Phospho (79.96633 Da) Charge: +2. Identified with: Sequest HT (v1.17); XCorr:4.52, ptmRS: best site probabilities: S14(Phospho) 100%. (B) Example results from label-free analysis of phosphopeptide enriched samples comparing 3 control versus 3 test samples. A two fold increase in TVLESGGEAGDGDS*DTTKK phosphopeptide expression in the sample compared to the control is observed.

4. Anticipated results

An example of a differentially expressed phosphopeptide from label-free analysis from H/ACA ribonucleoprotein complex subunit 4 is shown in Fig. 12.10.

An example of a differentially expressed phosphopeptide from TMT analysis from H/ACA ribonucleoprotein complex subunit 4 is shown in Fig. 12.11.

5. Optimization and troubleshooting/notes

- Note 1. Avoid excessive foaming of the sample when sonicating by allowing the sample to cool between sonication pulses or by reducing the length of the sonication pulses.
- Note 2. Reduction and alkylation using DTT and iodoacetamide of cysteine residues will minimize the appearance of unknown masses from disulfide bond formation and side-chain modification, improving the detection of cysteine-containing peptides.

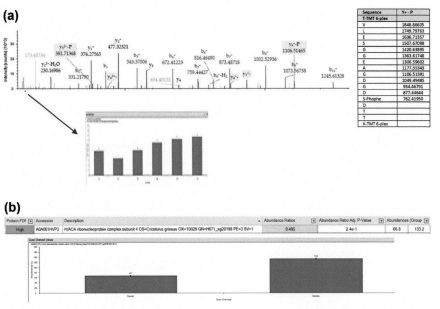

FIGURE 12.11

(A) Phosphopeptide from H/ACA ribonucleoprotein complex subunit 4 confidently identified following TMT labeling, phosphopeptide enrichment, and MS2 method. Sequence: TVLESGGEAGDGDSDTTK, S14-Phospho (79.96633 Da), K18-TMT6plex (229.16293 Da), T1-TMT6plex (229.16293 Da) Charge: +3, monoisotopic m/z: 759.69183 Da identified with: Sequest HT (v1.17); XCorr:4.02, ptmRS: best site probabilities: S14(Phospho): 97.18%. (B) Example results from TMT analysis of phosphopeptide enriched samples comparing 3 control versus 3 test samples pooled.

- Note 3. Alkylation of cysteines with iodoacetamide will increase the mass of a peptide by 57.02 Da for every cysteine present.
- Note 4. Although Lys-C is active in up to 6 M urea, trypsin is not active above 1.2 M urea.
- Note 5. An alternative approach to using Lys-C followed by trypsin is to use a double trypsin digestion approach initially at 1:100 enzyme: protein ratio for 4 h at 37°C followed by the addition of the same amount of trypsin again and incubating at 37°C overnight.
- Note 6. Retain the peptide sample flowthrough as a precaution, in case the peptide sample does not completely bind to the Sep Pak C18 column.
- Note 7. The phosphopeptide elution buffer has a high pH which can cause the loss of phosphates from phosphopeptides, therefore the sample needs to be evaporated as quickly as possible.
- Note 8. Avoid stainless steel emitter tips when sample ionization as the charged stainless-steel tip can irreversibly bind phosphopeptides.
- Note 9. SPS analysis of TMT samples provides more accurate quantitation of reporter ions compared to the MS2 (high resolution scanning method); however,

the method will provide less MS/MS information for the duration of the same LC separation as the MS instrument is carrying out an additional MS3 scan of the reporter region. It may be worth considering a longer separation time to compensate for this shortcoming.

6. Safety considerations

The techniques described here use biological materials and chemicals. Personal protection equipment must be worn at all times. When working with chemicals you should be very familiar with their Material Safety Data Sheets ahead of use. Validated biological safety cabinets and fume hoods should only be used. All waste generated (biological and chemical) should be disposed of in accordance with local guidelines and procedures (Figs. 12.10 and 12.11).

7. Summary

In this chapter, we have provided methodologies for protein digestion for large cell pellets for phosphopeptide enrichment followed by label-free analysis and for small cell pellets for TMT-label phosphopeptide enrichment.

Acknowledgments

This work was supported by a Science Foundation Ireland (SFI) Frontiers for the Future Award (grant no. 19/FPP/6759).

References

[1] Schelletter L, Albaum S, Walter S, Noll T, Hoffrogge R. Clonal variations in CHO IGF signaling investigated by SILAC-based phosphoproteomics and LFQ-MS. Appl Microbiol Biotechnol 2019;103(19):8127–43.

[2] Sharma K, D'Souza RC, Tyanova S, Schaab C, Wiśniewski JR, Cox J, et al. Ultradeep human phosphoproteome reveals a distinct regulatory nature of Tyr and Ser/Thr-based signaling. Cell Rep September 11, 2014;8(5):1583–94. https://doi.org/10.1016/j.celrep.2014.07.036. Epub 2014 Aug 21. PMID: 25159151.

[3] Walsh G. Biopharmaceutical benchmarks 2018. Nat Biotechnol Dec. 2018;36(12): 1136–45.

[4] Yusufi FNK, Lakshmanan M, Ho YS, Loo BLW, Ariyaratne P, et al. Mammalian systems biotechnology reveals global cellular adaptations in a recombinant CHO cell line. Cell Syst May 24, 2017;4(5):530–542.e6. https://doi.org/10.1016/j.cels.2017.04.009. PMID: 28544881.

[5] Dahodwala H, Kaushik P, Tejwani V, Kuo C-C, Menard P, et al. Increased mAb production in amplified CHO cell lines is associated with increased interaction of CREB1 with

transgene promoter. Curr Res Biotechnol 2019;1:49–57. https://doi.org/10.1016/j.crbiot.2019.09.001.

[6] Henry M, Power M, Kaushik P, Coleman O, Clynes M, Meleady P. Differential phospho-proteomic analysis of recombinant Chinese hamster ovary cells following temperature shift. J Proteome Res July 7, 2017;16(7):2339–58. https://doi.org/10.1021/acs.jproteo-me.6b00868. Epub 2017 Jun 6. PMID: 28509555.

[7] Kaushik P, Curell RVB, Henry M, Barron N, Meleady P. LC-MS/MS-based quantitative proteomic and phosphoproteomic analysis of CHO-K1 cells adapted to growth in glutamine-free media. Biotechnol Lett 2020;42:2523–36. https://doi.org/10.1007/s10529-020-02953-7.

[8] Taus T, Köcher T, Pichler P, Paschke C, Schmidt A, Henrich C, et al. Universal and confi-dent phosphorylation site localization using phosphoRS. J Proteome Res December 2, 2011;10(12):5354–62. https://doi.org/10.1021/pr200611n. Epub 2011 Nov 10. PMID: 22073976.

Ubiquitin diGly peptide enrichment and LC-MS/MS analysis to characterize the ubiquitinated proteome of mammalian cells

13

Karuppuchamy Selvaprakash[1], Michael Henry[1] and Paula Meleady[1,2]

[1]*National Institute for Cellular Biotechnology, Dublin City University, Glasnevin, Dublin, Ireland;*
[2]*School of Biotechnology, Dublin City University, Glasnevin, Dublin, Ireland*

1. Introduction

There are different types of PTMs that can modify the functionality of a protein including proteolytic cleavage and the covalent addition of a functional group to a protein. The covalent addition of a functional group includes modifications to the protein such as phosphorylation, glycosylation, acetylation, methylation, ubiquitination, lipidation, etc. Each type of PTM can alter the chemical and physical properties of proteins, thereby influencing their structure, stability, localization, and interaction with other molecules [1,2]. The identification and understanding of the role of PTMs in normal cellular biology and disease states is crucial for treatment and prevention of diseases. Ubiquitin modification is one of the most important PTMs and involves the covalent attachment of a protein called ubiquitin to a lysine residue of a target protein [3]. This modification is catalyzed by a series of enzymes, including ubiquitin-activating enzymes (E1), ubiquitin-conjugating enzymes (E2), and ubiquitin ligases (E3) [3,4]. Ubiquitination can have several effects on the target protein, including degradation by the proteasome, altering its subcellular localization, and regulating its activity, stability, and interaction with other proteins [4,5]. It is also becoming increasingly evident that ubiquitination events play a central role in normal physiology and disease progression [6].

Mass spectrometry has become an indispensable tool for the identification and characterization of proteins, including those with PTMs such as ubiquitination [7–9]. However, the relatively low stoichiometry of peptides bearing PTMs compared to their unmodified counterparts presents a technical challenge, and biochemical enrichment steps are generally necessary prior to mass spectrometry analysis. Due to the multifaceted roles of protein ubiquitination in mammalian cells

Proteomics Mass Spectrometry Methods. https://doi.org/10.1016/B978-0-323-90395-0.00006-1

and tissues, there is a great demand for the development of efficient analytical methods for the detection and quantification of ubiquitination sites on proteins [10]. A Lys-ε-diglycine (diGly) mark on peptides after tryptic digestion of proteins indicates a site of ubiquitination. Immunoaffinity-based enrichment of diGly remnant-containing peptides (using an anti-K-ε-GG antibody) is now one of the methods of choice for enrichment of ubiquitinated peptides for LC-MS/MS [11].

In this chapter, we describe a step-by-step comprehensive protocol for ubiquitinated peptide enrichment using an antibody-based immunoprecipitation of diGly peptides from a mammalian cell line (Chinese hamster ovary), followed by LC-MS analyses to detect and characterize the ubiquitinated proteome. The methods described are also applicable to differential ubiquitinated proteome studies and other mammalian cell lines (e.g., cancer cell lines).

2. Materials and equipment

- Phosphate Buffered Saline (PBS) buffer (e.g., Sigma/Merck #P5119).
- Urea (e.g., Sigma/Merck #U5128).
- NaCl.
- Tris base.
- Quick Start Bradford Protein Assay (Bio-rad # 5000201).
- DL-Dithiothreitol (DTT) (e.g., Sigma/Merck #D9779).
- Iodoacetamide (IAA) (e.g., Sigma/Merck #I1149).
- Sequencing-grade modified trypsin (e.g., Promega #V511A, Pierce/Thermo Fisher Scientific #90058).
- ProteaseMax Surfactant Trypsin Enhancer (Promega, #V2071).
- LysC protease enzyme (Pierce Lys-C Protease, #90307).
- Trifluoroacetic acid (TFA).
- Acetonitrile (ACN).
- 0.1% Formic Acid (FA).
- Ammonium Bicarbonate.
- Potassium Phosphate Monobasic (KH_2PO4).
- Potassium Chloride (KCl).
- SepPak tC18 reverse phase column (Waters—catalog #WAT036815).
- PTMScan HS Ubiquitin/SUMO Remnant Motif (K-ε-GG) Kit: (Cell Signaling Technologies, catalog #59322).
- PTMScan HS IAP Bind Buffer #1 (1X), contains 5% ACN (Cell Signaling Technology, #25144).
- PTMScan HS IAP Wash Buffer (1X) (Cell Signaling Technology, #42424).
- Sonicating probe.
- pH meter.
- 37°C CO_2 incubator.
- Refrigerated microcentrifuge.

- Microplate Spectrophotometer reading 96-well plate format at 660 nm (e.g., Multiskan GO, Thermo Scientific).
- End-over-end rotator.
- Vacuum Evaporator/Lyophilizer (e.g., SpeedVac, Thermo Scientific). NanoDrop One spectrophotometer (Thermo Fisher Scientific)
- Ultimate 3000 RSLC nano LC system (Thermo Scientific).
- Orbitrap Fusion Tribrid Mass Spectrometer (Thermo Scientific).
- Xcalibur software, version 2.0.7 (Thermo Scientific).
- Proteome Discoverer 2.5 (Thermo Scientific) with SEQUEST HT search algorithm and suitable latest *Cricetulus griseus* protein database in fasta format (see Note 1).

3. Before you begin

3.1 Solvent preparation

A number of solvents are required for the protocol. Ensure that all solvents and water that are used for all parts of the experiment are HPLC grade.

Prepare the following solvents using HPLC grade water.

1. 50% acetonitrile.
2. 0.1% trifluoroacetic acid.
3. 0.1% formic acid.

3.2 Sample preparation and proteolytic digestion

The following solutions are required for cell lysis and proteolytic digestion prior to mass spectrometry analysis.

1. 8 m urea: Prepare in HPLC grade water. Store aliquots for 6 months at −20°C.
2. 100 mM NaCl.
3. 50 mM Tris-HCl, pH 8.2.
4. 1% ProteaseMax: Add 100 µL of 100 mM ammonium bicarbonate to a new vial of ProteaseMAX surfactant for a 1% solution. Store aliquots at −20°C.
5. Lysis Buffer composition: 8 m Urea, 100 mM NaCl, 50 mM Tris-HCl, pH 8.2, 0.1% ProteaseMax (final concentration in the lysis buffer).
6. 50 mM ammonium bicarbonate: To prepare 50 mL of 100 mM solution, add 0.395 g and bring to 50 mL with LC-MS water. Prepare fresh on the day of use.
7. 0.5 M Dithiothreitol (DTT) solution: Prepare in 50 mM ammonium bicarbonate solution. Prepare shortly before use.
8. 50 mM Iodoacetamide solution: Prepare in 50 mM ammonium bicarbonate. Prepare shortly before use.

9. LysC protease enzyme (Pierce Lys-C Protease): Resuspend the contents of a 100 μg vial in 100 μL of 50 mM ammonium bicarbonate for a final concentration of 1 μg/μL before the digestion step.
10. Trypsin protease enzyme (Pierce Trypsin Protease): Prepare the stock concentration of 1 μg/μL using 50 mM ammonium bicarbonate before the digestion step.

4. Step-by-step method details

4.1 Cell harvest and lysis

1. Grow ~1×10^8 recombinant Chinese hamster ovary (CHO) cells to generate 5–10 mg of soluble protein. See Note 2.
2. Harvest cells by centrifugation at $300 \times g$ for 5 min at room temperature.
3. Carefully remove the supernatant and wash with prechilled PBS buffer three times by repeating steps 2 and 3.
4. Store the cell pellet at −80°C until cell lysis is performed.
5. Lyse the cell pellets with 1.5–2 mL of lysis buffer. Resuspend the cell pellet in the lysis buffer and pipette the slurry up and down well.
6. Disrupt the cells using a sonicating probe while maintaining the lysate at 4°C.
7. Centrifuge the sample at 14,000 rpm for 15 min at 4°C using a refrigerated microcentrifuge. Transfer the supernatant into a fresh microcentrifuge tube.
8. Determine the protein concentration using the Quick Start Bradford Protein Assay according to manufacturer's instructions, or alternative suitable protein assay.

4.2 In solution protein digestion

1. Prior to digestion, dilute the sample with 50 mM of freshly prepared ammonium bicarbonate to achieve a final urea concentration of <1.2 M.
2. Add 0.5 M DTT and heat the lysate at 56°C for 20 min to reduce the disulfide bonds of the proteins.
3. Add 50 mM iodoacetamide and incubate at room temperature in the dark for 1 h for alkylation.
4. Add Lys-C solution to the sample at 1:200 (w/w) (protein:enzyme) and allow to digest at 37°C for 4 h.
5. Add sequence-grade trypsin at 1:100 (w/w) (protein:enzyme) and allow to digest overnight at 37°C.
6. Add TFA to a final concentration of 0.1% to inactivate the trypsin.
7. Proceed directly to desalting or store the sample at −80°C until required.

4.3 **Peptide desalting**

1. Acidify the sample for efficient peptide bind by adding TFA with a final concentration of 0.5%. Check the pH by spotting a small amount of peptide sample on a pH strip.
2. Prewet the C18 column with 5 mL of 100% ACN. See Note 3.
3. Wash the C18 column with 4 mL of 50% ACN and 0.1% FA to clear any unwanted material bound to the cartridges.
4. Equilibrate with 1 mL of 0.1% FA. Carry this step out three times.
5. Load the acidified cell lysate digest to the C18 column.
6. Wash and desalt with 4 mL of 0.1% FA.
7. Elute the peptides bound to the C18 cartridges twice with 2 mL of 50% ACN, 0.1% FA.
8. Freeze the eluate by storing at $-80°C$ for ~ 4 h. Lyophilize for a minimum 1 day to remove all the residual acid from the peptide sample. The peptides should be completely dry prior to diGly enrichment.

4.4 **Ubiquitin diGly peptide enrichment**

1. Dissolve the dry peptides in 1.5 mL of HS IAP bind buffer (1X) (approximately 5 mg of peptide is recommended for the experiment). Resuspend the sample mechanically by gently pipetting up and down.
2. Determine the peptide concentration of the sample using a NanoDrop One.
3. After dissolving the peptide, check the pH of the peptide solution by spotting a small volume on pH indicator paper. The pH should be close to neutral (should not be lower than 7.0). If necessary, add 2 μL of 1 m Tris base until the pH is at ~ 7.
4. Clear the solution by centrifugation for 5 min at $10,000 \times$ g at 4°C. Cool on ice.
5. Spin down the vial of antibody-bead slurry at $2000 \times$ g for ~ 1 min to bring down any of the beads clinging to the sides and cap of the stored vial.
6. Transfer 20 μL of bead slurry to a 1.5 mL microcentrifuge tube and wash with 1 mL of ice cold PBS by inverting the tube for ~ 5 times.
7. Centrifuge the vial at $2000 \times$ g and remove the supernatant.
8. Add the soluble peptide solution into the tube containing the antibody beads and pipette the mixture gently.
9. Tighten the cap and seal the top of the tube with Teflon tape to avoid any leakage.
10. Incubate the mixture on an end-over-end rotator for 2 h at 4°C.
11. After 2 h, spin down the tube at $2000 \times$ g for 1 min to bring down the beads.
12. Wash the beads by adding 1 mL of HS IAP wash buffer (1X) and centrifuge at $2000 \times$ g for 1 min. Repeat this step an additional 3 times.
13. Wash the beads further by adding 1 mL of ice-cold LC-MS water and centrifuge at $2000 \times$ g for 1 min. Repeat this step an additional 2 times.

14. Elute the diGly peptides by adding 50 μL of HS IAP elution buffer to the beads and incubate for 5 min at room temperature.
15. Mix the beads gently every 2 min.
16. Centrifuge the sample at 2000 × g for 1 min and collect the supernatant.
17. Repeat the elution step with an additional 50 μL of IAP elution buffer.
18. Proceed with peptide desalting as described in Section 4.3.

4.5 LC-MS/MS analysis

1. Resuspend the diGly peptide samples in 20 μL of LC-MS grade water with 0.1% FA and 2% ACN.
2. Carry out nano LC−MS/MS analysis using, for example, an Ultimate 3000 RSLCnano system coupled to an Orbitrap Fusion Tribrid Mass Spectrometer.
3. Load the sample onto a C18 trap column.
4. Desalt the sample for 3 min using a flow rate of 25 μL/min in 0.1% TFA containing 2% ACN (Loading Buffer).
5. Switch the trap column online with the analytical column using a column oven at 45°C and elute the peptides with the following binary gradients of: Mobile Phase Buffer A and Mobile phase buffer B: 0%−25% solvent B in 120 min and 25%−50% solvent B in a further 60 min, where solvent A consisted of 2% acetonitrile (ACN) and 0.1% formic acid in water and solvent B consisted of 80% ACN and 0.08% formic acid in water.
6. Set the column flow rate to 300 nL/min.
7. Operate the Orbitrap Fusion Tribrid Mass Spectrometer with a resolution of 120,000 (at m/z 200), a maximum injection time of 90 ms and an automatic gain control (AGC) value of 4×10^5 to perform full MS scans for peptide mass information. Carry out data-dependent acquisition using a full scan range of 380−1500 m/z. Use a top-speed MS/MS acquisition algorithm to determine the number of selected precursor ions for fragmentation. A dynamic exclusion is applied to analyzed peptides after 60 s to avoid repeat fragmentation of the same peptide and only peptides with a charge state between 2+ and 7+ are analyzed. The peptides are fragmented using higher energy collision-induced dissociation (HCD) with a normalized collision energy of 28%. The resulting MS/MS fragment ions are measured in the linear ion trap.

4.6 Peptide and protein identification

1. Search the raw MS/MS data using Proteome Discoverer 2.5 against *Cricetulus griseus* UniProt database using the SEQUEST HT algorithm (see Note 1).
2. Apply the following search parameters for all MS2 data for protein identification: (i) set 20 ppm precursor ion tolerance and 0.6 Da fragment ion tolerance; set lysine with a diGly remnant (+114.04 Da), oxidation of methionine (15.99491 Da) and N-terminal acetylation (42.010564 Da) as variable modifications. Set carbamidomethylation of cysteine (125.047679 Da) as a fixed modification. Filter the peptide matches to a peptide false discovery rate of <1%.

5. Expected outcomes

Fig. 13.1 shows an example of a typical LC chromatogram of enriched diGly peptides sample from mammalian cells. The most abundant peptide, LIFAGK(GG) QLEDGR (m/z 730.9), in the enriched sample is the K48 modified tryptic diGly peptide of ubiquitin. If this peak is absent from the chromatogram, diGly peptides enrichment is most likely unsuccessful.

6. Optimization and troubleshooting/notes

Note 1: The cell line used in this protocol is from Chinese hamster ovary. We therefore regularly download a FASTA database from the reference proteome for *Cricetulus griseus*. For human cells, a FASTA database for the Homo sapiens reference proteome should be regularly downloaded. Refer to https://www.uniprot.org/uniprotkb?query=Cricetulus+griseus

Note 2: A large number of cells are required for ubiquitination enrichment experiments using this method. The number of cells needed to yield a protein concentration between 5 and 10 mg may need to be optimized for the cell line used.

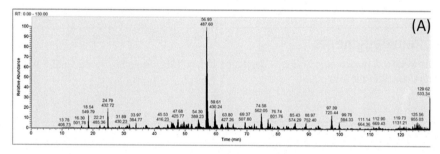

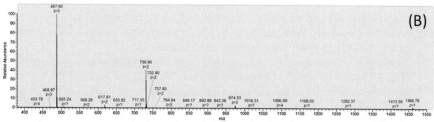

FIGURE 13.1

(A) LC chromatogram of diGly peptides sample from CHO-DP12 cells separated using a 120 min gradient. (B) The most abundant peptide LIFAGK(GG)QLEDGR (m/z 730.9) in the enriched sample is the K48 modified tryptic diGly peptide of ubiquitin. If this peak is absent from the chromatogram, diGly peptides enrichment is most likely unsuccessful.

Note 3: The application of all solutions during the desalting procedure should be performed by gravitational flow.

7. Safety considerations and standards

The techniques described here use biological materials and chemicals. Personal protection equipment must be worn at all times. When working with chemicals you should be very familiar with their Material Safety Data Sheets ahead of use. Validated biological safety cabinets and fume hoods should only be used. All waste generated (biological and chemical) should be disposed of in accordance with local guidelines and procedures.

8. Summary

In this chapter, we have provided a methodology for enrichment of ubiquitinated peptides from a mammalian cell line using an immunoprecipitation approach, and carried out using commercially available kit. The method can be easily adapted to other cell lines and tissue samples allowing the ubiquitinated proteome to be studied in detail.

Acknowledgments

This work was funded by a Science Foundation Ireland (SFI) Frontiers for the Future Award (grant no. 19/FPP/6759).

References

[1] Deribe YL, Pawson T, Dikic I. Post-translational modifications in signal integration. Nat Struct Mol Biol 2010;17:666−72. https://doi.org/10.1038/nsmb.1842.

[2] Mann M, Jensen ON. Proteomic analysis of post-translational modifications. Nat Biotechnol 2003;21:255−61. https://doi.org/10.1038/nbt0303-255.

[3] Pickart CM. Mechanisms underlying ubiquitination. Annu Rev Biochem 2001;70:503−33. https://doi.org/10.1146/annurev.biochem.70.1.503.

[4] Neutzner M, Neutzner A. Enzymes of ubiquitination and deubiquitination. Essays Biochem 2012;52:37−50. https://doi.org/10.1042/bse0520037.

[5] Peng J, Schwartz D, Elias JE, Thoreen CC, Cheng D, et al. A proteomics approach to understanding protein ubiquitination. Nat Biotechnol 2003;21:921−6. https://doi.org/10.1038/nbt849.

[6] Steger M, Karayel Ö, Demichev V. Ubiquitinomics: history, methods, and applications in basic research and drug discovery. Proteomics 2022;22:2200074. https://doi.org/10.1002/pmic.202200074.

[7] Danielsen JMR, Sylvestersen KB, Bekker-Jensen S, Szklarczyk D, Poulsen JW, et al. Mass spectrometric analysis of lysine ubiquitylation reveals promiscuity at site level. Mol Cell Proteomics 2011;10:003590. https://doi.org/10.1074/mcp.M110.003590.

[8] Kaiser P, Wohlschlegel J. Identification of ubiquitination sites and determination of ubiquitin-chain architectures by mass spectrometry. In: Methods in enzymology. Elsevier; 2005. p. 266−77. https://doi.org/10.1016/S0076-6879(05)99018-6.

[9] Udeshi ND, Mertins P, Svinkina T, Carr SA. Large-scale identification of ubiquitination sites by mass spectrometry. Nat Protoc 2013;8:1950−60. https://doi.org/10.1038/nprot.2013.120.

[10] Sahu I, Zhu H, Buhrlage SJ, Marto JA. Proteomic approaches to study ubiquitinomics. Biochim Biophys Acta Gene Regul Mech 2023;1866:194940. https://doi.org/10.1016/j.bbagrm.2023.194940.

[11] Van Der Wal L, Bezstarosti K, Sap KA, Dekkers DHW, Rijkers E, et al. Improvement of ubiquitylation site detection by orbitrap mass spectrometry. J Proteonomics 2018;172:49−56. https://doi.org/10.1016/j.jprot.2017.10.014.

Liquid chromatography mass spectrometry —based proteomics: Global cell proteome profile

14

Giorgio Oliviero[1], Kieran Wynne[1] and Paula Meleady[2,3]

[1]*Systems Biology Ireland, School of Medicine, University College Dublin, Belfield, Dublin, Ireland;*
[2]*School of Biotechnology, Dublin City University, Glasnevin, Dublin, Ireland;* [3]*National Institute
for Cellular Biotechnology, Dublin City University, Glasnevin, Dublin, Ireland*

1. Introduction

Mass spectrometry—based proteomics refers to large-scale studies of the "proteome" employing a combination of liquid chromatography (LC) coupled with tandem mass spectrometry (LC-MS/MS) [1].

The term "proteome" denotes the entire set of proteins expressed by an organism or the set of detectable proteins in a nonliving sample (blood, tissue, etc.) [2,3].

Two main modes of approach exist: "top-down" proteomics, where intact protein molecules are studied, and "bottom up" or shotgun proteomics, where the proteins are first processed using endopeptidases to facilitate downstream analysis [4,5].

Combined improvements in separation technology, mass spectrometry, and data analysis software now permit the identification and quantitation of thousands of peptides using the shotgun approach in a single study. The ability to simultaneously report thousands of proteins allows integration with other "omics" platforms (e.g., transcriptomics, metabolomics), and can lead to powerful new biological insights into disease.

In clinical applications, mass spectrometry analytical strategies can be applied to acquire valuable information describing the molecular mechanisms of tumors, for example, to enable better classification of different cancer types [6].

The protein complement reflects the phenotypic consequence of disease, allowing a connection between the relatively static genetic information with the dynamic proteomic landscape within the cell to be established [6]. Furthermore, clinical proteomics offer a link to potential pharmacological intervention since the majority of current druggable targets in tumor cells are proteins.

However, the clinical environment presents challenges, since valuable and often small amounts of clinical material may only be available. To detect differentially expressed proteins in a clinical context, specialized proteomics techniques requiring significant technical expertise and knowledge are needed. The key steps in achieving a successful experiment that need to be optimized include (1) the ability to analyze many (possibly hundreds) of samples in an uninterrupted fashion to achieve sufficient statistical power in patient cohorts; (2) establishing automated workflows with minimal manual handling; (3) implementing longitudinal standardization procedures; (4) shortening experiment timelines; (5) promoting cost-effectiveness for analytical laboratories. Simultaneously achieving optimal performance across all these objectives is an enormous challenge, yet they are essential factors for a successful procedure [7].

Despite vast technical improvements, MS-based proteomics is not yet routinely established in a clinical environment [6]. Most of these bottlenecks can be resolved by implementing automation procedures that avoid manual handling, and thereby eliminate the risk of error and variability. A number of automation tools permitting parallelized sample preparation are commonly used in clinical proteomic research [8–13]. This provides an attractive and cost-effective solution for routine and comprehensive clinical studies, easing the introduction of translational proteomic research with minimal hands-on time and low sample consumption.

Historically, in many proteomic applications for clinical research, the predominant mass spectrometry strategy employed label-free analysis using data-dependent acquisition (DDA). In DDA, the eluted peptides from a liquid chromatography system are detected by a first stage MS1 scan, generally within a fixed mass range. Subsequently, a fixed number of the most intense peptide ions are then selected for the second-stage tandem mass spectrometry (MS/MS) [6]. However, due to the semistochastic nature of precursor selection for MS/MS, each peptide may not be consistently detected and therefore may not be quantified in all samples, resulting in low proteome coverage arising from a large fraction of missing values in the dataset [6,14].

To overcome this problem, new strategies known as DIA-MS (data-independent acquisition) and SWATH-MS (sequential window acquisition of all theoretical mass spectra) provide better reproducibility and sensitivity when compared to conventional DDA-MS [15,16]. DIA-MS is based on the fragmentation of all precursor ions identified in the MS1 survey scan, where fragment ions are accumulated in a fixed number of wide isolation windows that span the entire mass-to-charge ratio (*m/z*) range [6,17]. DIA workflows were recently demonstrated to achieve higher proteomic depth in single injections than conventional DDA proteomic strategy [16,17]. DIA now routinely reaches protein coverage levels that are comparable to, or even exceed, that achieved by DDA [18,19].

However, the computational processing of DIA datasets remains challenging owing to their inherent complexity [20]. In order to meet these challenges, three data analysis strategies for DIA have recently been developed: (1) using spectral

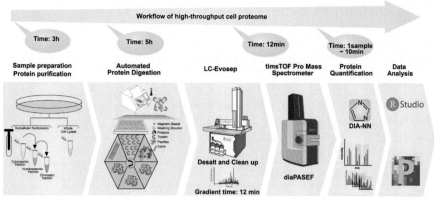

FIGURE 14.1

Automated quantitative proteomics strategy for cell proteome analysis. Following subcellular fractionation purification, proteins were analyzed using high-resolution quantitative mass spectrometry coupled with biochemistry and bioinformatics analysis to promote in-depth cell proteome coverage.

libraries generated by analyzing DDA runs of prefractionated samples (on-line and/or off-line 2D LC-MS/MS); (2) using spectral libraries generated by refining predicted libraries using gas-phase-fractionation (GPF); (3) using no additional experimental data to generate spectral libraries, but harnessing existing experimental data using advanced computing approaches (e.g., neural networks and deep learning) [16,19–23].

We describe an MS-based clinical proteomic strategy used to analyze proteins extracted from a tumor cell in a high throughput manner (Fig. 14.1). We first employed a subcellular fractionation strategy to promote in depth proteome coverage and gain insights that reflect the cancer cell context [24–26]. Then we applied an automated high throughput workflow using automated protein purification systems (KingFisher Duo Prime) combined with a chromatography system (Evosep One) that delivers fast, sensitive, and robust proteomic analysis of clinical samples in a less time-consuming manner [27].

For mass spectrometry analysis we employed DIA-PASEF, a workflow made possible using the Trapped Ion Mobility Separation (TIMS) device within the timsTOF Pro mass spectrometer (Bruker Daltonics) [28]. Our approach provides a significant improvement in the depth of proteome coverage and quantitative precision. Downstream of the experiment, data are deconvoluted through the use of deep neural networks in DIA-NN, an automated software that simplifies and accelerates DIA output analysis [21]. DIA-NN delivered fast and accurate in-depth proteome coverage [21].

2. Key resources table

Reagent or resource	Source	Identifier
Antibodies		
Histone H3	Abcam	ab1791
B23/Nucleophosmin	SantaCruz	271737
GAPDH	Cellsignal	2118
Beta-actin	Sigma Aldrich	A2228
Experimental models: cell lines		
NTera-2/cloneD1 (NT2) cells	ATCC	CRL-1973
Software and algorithms		
RStudio		Integrated development for R. RStudio, PBC, Boston, MA URL http://www.rstudio.com/.
Perseus		https://maxquant.net/perseus/

3. Materials and equipment

1. Corning tissue-culture treated culture dishes (Sigma−Aldrich CLS430599)
2. Corning 15 mL centrifuge tubes (Sigma−Aldrich CLS430791)
3. Eppendorf LoBind microcentrifuge tubes (Sigma−Aldrich EP0030108116)
4. Sterile Millipak-40 Filter Unit 0.22 µm (Millipore MPGL04GH2)
5. 2 mL Dounce homogenizer and type B pestle (Sigma−Aldrich D8938)
6. HPLC-grade H_2O (Fisher Scientific W/0112/17)
7. Micrococcal nuclease (MNase; Sigma−Aldrich N5386)
8. Ethylene glycol-bis(2-aminoethylether)-N,N,N′, N′-tetra acetic acid (EGTA; Sigma−Aldrich E3889)
9. Potassium chloride (KCl) for molecular biology, $\geq 99.0\%$ (Sigma−Aldrich P9541)
10. Sodium chloride (NaCl) for molecular biology, $\geq 99.0\%$ (Sigma−Aldrich S3014)
11. Magnesium chloride (MgCl) (Sigma−Aldrich M8266)
12. HEPES (4-(2-hydroxyethyl)-1-piperazineethanesulfonic acid) 1 M buffer solution (Sigma−Aldrich 83264)
13. Trizma hydrochloride (TRIS-HCl) (Sigma−Aldrich T5941)
14. Calcium chloride dehydrate ($CaCl_2 \cdot 2H_2O$) for molecular biology, $\geq 99.0\%$ (Sigma−Aldrich C3306)
15. UltraPure 0.5 M EDTA, pH 8.0 (Thermofisher 15575020)
16. HEPES (Sigma−Aldrich H4034)

17. cOmplete Mini, EDTA-free Protease Inhibitor Cocktail (Roche 11836170001)
18. PhosSTOP Phosphatase Inhibitor Cocktail Tablets (Roche 4906837001)
19. Phosphate-buffered saline (PBS) (Thermofisher 003002)
20. Dithiothreitol (DTT) (Sigma—Aldrich 10708984001)
21. Iodoacetamide (Sigma—Aldrich I6125)
22. Water (LC-MS Grade) (Sigma—Aldrich 1153331000)
23. Hydrochloric acid (HCl) for molecular biology (Sigma—Aldrich H1758)
24. Trifluoroacetic acid (TFA), sequencing grade (Thermofisher 28904)
25. Acetonitrile (ACN) hyper-grade for LC-MS (Sigma—Aldrich 1000291000)
26. Formic acid (FA) LC-MS grade (Fisher A117)
27. Acetic acid (Sigma—Aldrich 695092)
28. Sequencing grade modified Trypsin (Promega V5111)
29. Benzonase nuclease (Sigma—Aldrich E1014-25KU)
30. Sodium butyrate (Sigma—Aldrich B5887)
31. Deacetylase inhibitor (Active Motif 37494)
32. Ammonium bicarbonate (Sigma—Aldrich A6141)
33. Urea for molecular biology (Sigma—Aldrich U5378)
34. 660 nm Protein Assay Reagent (ThermoFisher 22660)
35. NuPAGE 10%, Bis-Tris, 1.0 mm, Mini Protein Gel, 10-well (ThermoFisher NP0301BOX)
36. NuPAGE MOPS SDS Running Buffer (20×) (ThermoFisher NP000102)
37. C18Tips (ThermoFisher 87784)
38. Vacuum concentrator (Speed-Vac)
39. Eppendorf microcentrifuge Refrigerated
40. Sonicator (Syclon SKL-150W)
41. Ethyl alcohol, Pure (Sigma—Aldrich E7023)
42. KingFisher Duo Prime Purification System (ThermoFisher 5400110)
43. KingFisher Deepwell 96 Plate, V-bottom (ThermoFisher 95040450)
44. KingFisher Deepwell 96 Plate tip comb for DW magnets (ThermoFisher 97002534)
45. Magnetic Speed Beads (Sigma—Aldrich GE65152105050250 and GE45152105050250)
46. DPBS (10×), no calcium, no magnesium (Invitrogen 14200075)
47. Trypsin 0.25% EDTA for tissue culture (Invitrogen 25200056)
48. Sodium butyrate (Sigma—Aldrich B5887)
49. AEBSF (Millipore Corp 101500)

4. Before you begin

4.1 Cell fractionation for mass spectrometry analysis

4.1.1 Cell lysis buffers

Prepare all buffers using LC-MS grade water. Prepare all buffers with Roche cOmplete EDTA-free protease inhibitor cocktail and precool to 4°C on ice.

1. Whole Cell Lysis Buffer: 25 mM TrisHCl pH 7.4, 150 mM NaCl, 2 mM EDTA, 0.2% NP-40, 0.1% SDS.
2. Hypotonic Buffer (Buffer A): 10 mM HEPES pH 7.9, 1.5 mM $MgCl_2$, 10 mM KCl, 0.5 mM DTT.
3. Buffer MNase (Buffer B1): 10 mM Tris-HCl pH 7.4, 2 mM $MgCl_2$, 5 mM $CaCl_2$.
4. Buffer MNase Wash (Buffer B2): 10 mM Tris-HCl pH 7.4, 2 mM $MgCl_2$.
5. Salt Buffer Base (Buffer C): 10 mM Tris-HCl pH 7.4, 2 mM $MgCl_2$, 2 mM EGTA (pH ~7.0–8.0), 1 mM EDTA pH 10.0, NaCl variable concentration from 50 to 600 mM (see Note 1).

4.2 Protein extraction buffers for mass spectrometry
4.2.1 Mass spectrometry buffers

1. Urea Buffer (8 M): Resuspend one tube containing 4.8 g and LC-MS grade water till 10 mL. Mix well until the urea dissolves completely into solution (see Note 2).
2. Ammonium Bicarbonate (100 mM): Resuspend one tube containing 79 mg with 10 mL of LC-MS grade water.
3. Calcium Chloride (100 mM): Resuspend one tube containing 147 mg with 10 mL of LC-MS grade water.
4. Dithiothreitol: Make 1.25 M stock. Resuspend one tube containing 192.8 mg with 1 mL LC-MS grade water. Divide into 25 μL aliquots. Store at −20°C for up to 1 year. Thaw one aliquot for each experiment.
5. Iodoacetamide solution: Make 100 mM stock. Weigh out 19 mg of iodoacetamide and cover the tube with foil to protect it from light. Dissolve the powder in LC-MS grade water to a final volume of 1 mL immediately before use (see Note 3).
6. Trypsin: Resuspend trypsin at 1 μg/μL. Dissolve 20 μg trypsin in 40 μL with 50 mM ammonium bicarbonate prior to use. Aliquot and store at −80°C.
7. 20% TFA: Add 20 mL TFA to 80 mL LC-MS grade water. Store at room temperature.

4.2.2 Automated protein digestion solutions

1. Pure Ethanol 80%: add 40 mL of pure ethanol to 10 mL of LC-MS grade water. Store at room temperature.
2. Ammonium Bicarbonate (50 mM): resuspend one tube containing 39.5 mg with 10 mL of LC-MS grade water.

5. Step-by-step method details
5.1 Collection of cells from culture

1. If cells are grown in suspension, collect cells by centrifugation at 300 × g for 5 min. If adherent, aspirate and discard the cell medium. Rinse the attached cells with PBS without Ca^{2+} and Mg^{2+} (see Note 4).

2. Incubate the cells in either trypsin or trypsin-EDTA with enough volume to cover the surface of the plates at 37°C until the cells detach (time varies for different cell lines).
3. Collect cells by centrifugation at 300 × g for 5 min at 4°C.
4. Wash cells two more times in cold PBS and carefully aspirate off all the PBS.
5. The cell pellet can then be used immediately or else snap-frozen in liquid nitrogen and stored at −80°C until ready to process.

5.1.1 Whole-cell lysate protein isolation
Time: 45 min.

1. Thaw the cell pellet on ice until it is loose (∼10 min).
2. Estimate the pcv (packed cell volume) and resuspend the pellet in cold PBS (see Note 5).
3. Gently resuspend the cells in PBS and transfer to a new 1.5 mL low-binding tube.
4. Centrifuge at 300 × g for 5 min at 4°C and remove the supernatant.
5. Resuspend the pellet in Whole Cell Lysate buffer.
6. Incubate the sample on ice for 15 min.
7. Sonicate samples at 5−15 W output power 3 × 10 s each, cooling on ice for 1 min between each burst (see Note 6).
8. Preclear lysates by spinning at max speed for 15 min at 4°C.
9. Remove supernatant and transfer to a new 1.5 mL microcentrifuge tube.

5.1.2 Isolation of intact nuclei
Time: 60 min.

1. Thaw the cell pellet on ice until it is loose (∼10 min).
2. Estimate the pcv (packed cell volume) and resuspend the pellet in 2−3 times *pcv* with Buffer A.
3. Gently resuspend the cells in Buffer A and transfer to a new 1.5 mL low-binding tube (see Note 7).
4. Incubate the sample on ice for 30 min.
5. Transfer the sample to the prechilled Dounce homogenizer.
6. Break the cell membrane by 40 up/down strokes (see Note 8).
7. Transfer the sample to a new tube.
8. Centrifuge at 500 × g for 10 min at 4°C. If all nuclei are pelleted and no intact cells remain, after checking under microscope, then proceed to step 10 (see Note 9).
9. Move supernatant carefully to a new tube labeled "cytosol fraction."
10. Add 1 mL of Buffer A and resuspend the nuclei pellet to wash away residual cytosolic proteins.
11. Centrifuge for 10 min as in step 8 and discard the supernatant.
12. The remaining pellet is the intact nuclei.

5.1.3 Complete DNA digestion and salt-extraction of chromatin-associated proteins
Time = 120 min.

1. Estimate the pcv (packed cell volume) and gently resuspend the pellet 2–3 times pcv with Buffer B1.
2. Warm sample to 37°C for 5 min before adding MNase.
3. Perform a complete MNase digestion of the nuclei by MNase to a final concentration of 1 U/µL per reaction.
4. Incubate at 37°C for 30 min, with gentle rocking to prevent aggregation of the nuclei.
5. Stop the MNase digestion by adding EGTA solution to a final concentration of 2 mM per reaction. Gently mix the sample with a pipette.
6. Centrifuge the sample at 500 × g for 10 min at 4°C.
7. Carefully transfer the supernatant to a new tube labeled "MNase fraction." Sample can be stored at −80°C until ready to process (see Note 10).
8. Wash the nuclei pellet with 400 µL of Buffer B2 by resuspending the pellet.
9. Centrifuge the sample at 500 × g for 10 min at 4°C and discard the supernatant.
10. Gently resuspend the MNase-digested nuclei in 400 µL of cold Salt Buffer C.
11. Add Benzonase Nuclease to a final concentration of 0.15 U/µL and incubate for 90 min at 4°C (see Note 11).
12. Centrifuge at 4°C, 10,000 × g for 20 min.
13. Retain the supernatant and transfer to new appropriately labeled tube.
14. The leftover pellet can be processed for histone purification extraction (see Note 12).
15. Fig. 14.2 shows an example of the successful protein lysate separation through western blot analysis.

5.2 Protein digestion into peptides
Time: 40 min.

1. Prior to protein denaturation measure the protein concentration of the samples using any protein quantification assay (see Note 13). Protein denaturation: gently resuspend the sample in urea (final concentration, 4 M), ammonium bicarbonate (final concentration, 100 mM), and calcium chloride (final concentration, 100 mM).
2. To reduce the cleared protein lysate samples, add DTT (final concentration, 1 mM), mix well and incubate at room temperature for 15 min.
3. Alkylate samples by adding iodoacetamide solution (final concentration, 3 mM), mixing well, and incubate at room temperature for 15 min in the dark.
4. Samples are ready for trypsin digestion.

5.2.1 "KingFisher" automated protein digestion system
Time 4 h and 30 min (Fig. 14.3).

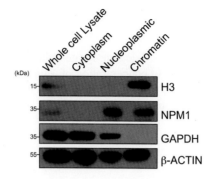

FIGURE 14.2

Western blot analysis of subcellular fractionation. The cytoplasmic, nucleoplasmic, and chromatin fractions are probed (respectively) by GAPDH, NPM1, and H3 antibodies. Beta-actin antibody was used as a loading control.

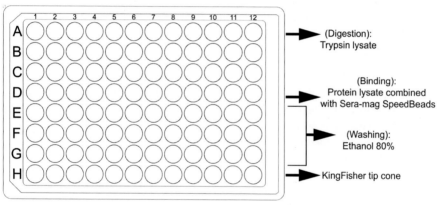

FIGURE 14.3

Design of the protein digestion plate.

1. Bring Sera-mag SpeedBeads to room temperature and gently vortex to fully resuspend magnetic beads.
2. Beads come at a stock concentration of 50 μg/μL. Take the required number of beads and mix hydrophilic and hydrophobic beads at a 1:1 ratio into a tube (see Note 14).
3. Place the tube on a magnetic rack and wait 30 s for the liquid to clear. Discard the supernatant.
4. Remove the tube from the magnetic rack and add 1 mL of water and bring the beads into solution by flicking the tube.
5. Place on the magnetic rack again, wait for the liquid to clear (approx. 30 s) and discard the supernatant. Repeat the wash step twice.

6. Resuspend the beads in 100% pure ethanol and transfer samples into the plate, lane D.
7. Binding [8]:
   i. Lysate volume (V_{Lysate}) = Calculate lysate volume to reach desired protein amount, keeping a protein concentration of 1 μg/μL.
   ii. 100% EtOH volume = V_{Lysate} (to reach 50% EtOH v/v).
   iii. We recommend using 1 μL of 10 μg/μL of beads mix per μg of protein to be processed.
8. Automated trypsin digestion setting:
   i. incubation time 4 h at 37°C.
   ii. mix method: low speed.

5.2.2 Terminate protein digestion

1. Once the protein digestion has terminated, remove the samples from Lane A and transfer them into a low binding tube (see Note 15).
2. Acidify the samples with 100% TFA to achieve a final concentration of 0.1% −1% TFA, and vortex the samples. Peptides are dried in a Speed-vac and stored at −20°C until MS analysis.

5.2.3 Manual sample desalting

1. After protein digestion, salt might be present in the sample. It is necessary to remove the excess salt to correctly identify/quantify the peptides. Also, salts impede HPLC-MS analysis because they ionize during electrospray, suppressing the signal from peptides.
2. Sidoli et al. described desalting samples though C18 Stage-tips [29].
3. Alternatively, peptides are desalted, cleaned, and concentrated on C18 Tips according to the manufacturer's instructions [30].

5.3 Combination of automated sample desalting and mass spectrometry analysis

1. Samples are redissolved in 0.1% TFA/2.5% acetic acid in H_2O to reach a final concentration of 1 μg/μL.
2. Analysis is carried out on an Evosep One liquid chromatography system connected to a mass accuracy high resolution mass spectrometry, timsTof Pro (Bruker Daltonics).
3. For proteome analysis, peptides are loaded onto disposable Evosep tips. Then, the Evosep tips are placed in position on the Evosep One, in a 96-tip box. The autosampler is configured to pick up each tip, elute and separate the peptides over a performance column 8 cm × 150 μm, 1.5 μm C18, (EV1109, Evosep) using a set chromatography method (100 samples a day/total gradient time 11.5 min/cycle time 14.4 min per sample) [27,31].
4. Prior to creating a data-independent analysis (DIA) method, a pooled sample from the trypsin digested fractions is analyzed using data-dependent analysis

parallel accumulation serial fragmentation (DDA-PASEF). The settings are capillary voltage 1700 V, dry gas flow 3 L/min, and dry temperature 180°C. A scan range of (100−1700 *m/z*) is performed at a rate of five PASEF MS/MS frames to 1 MS scan with a cycle time of 1.03 s [32].

5. The dda-PASEF file is used to create the dia-PASEF method within Bruker timsControl software. The scan mode "dia-PASEF" is selected and the pooled dda-PASEF file is opened in the window editor in the MS/MS tab.

6. We use the adjustable rhomboid to select the area of a heatmap where the identifiable peptides (central region in the heatmap containing peptides with charge states from +2 to +5) can be found.

7. All data is acquired using data independent analysis parallel accumulation serial fragmentation (dia-PASEF).

8. In summary, the following DIA-PASEF settings are used; mass width 25.0 Da, mass overlap 0.0, mass steps per cycle 34, mobility overlap 0.00, mass range 366.7−1239.8 *m/z*, and mean cycle time estimate 3.96 s [28].

5.4 DIA data processing

1. The raw data is searched against *Homo sapiens*, Uniprot Swissprot database (reviewed, June 20, 2021) using the search engine developed in DIA-NN (version 1.8) [21].

2. DIA-NN is used to generate in silico *Homo sapiens* spectral library with the following criteria: enzyme is set to trypsin/P with up to one missed cleavage. Carbamidomethylation (C) and oxidation (M)/acetylation (protein N-term) are selected as a fixed and variable modification, respectively. Peptide lengths are set from 7 to 30 and precursor charge rates between 2 and 5.

3. DIA-NN is operated with maximum mass accuracy tolerances set to 10 ppm for both MS1 and MS2 spectra. The scan window is set at 0 to let DIA-NN perform automatic inference for the first run of the experiment.

4. The dataset is analyzed with match-between-runs (MBR) enabled, similarly to the previous dia-PASEF workflow [28]. Quantification mode is set to "Robust LC (high precision)." All other settings are default.

5. The DIA-NN's search outputs have a 1% FDR setting applied, with the following criteria: precursor q-value <1% and global protein q-value <1%. No additional normalization steps are performed, as the resulting LFQ intensities are normalized by the MaxLFQ [33] procedure. The protein group table is listed in the supplementary material.

5.5 Data interpretation

1. Bioinformatic analysis of the DIA-NN output file (PG, "protein group") is conducted by Perseus software (version 1.6) [34] and RStudio employing the following packages: ggplot2 and ggrepel (Fig. 14.4).

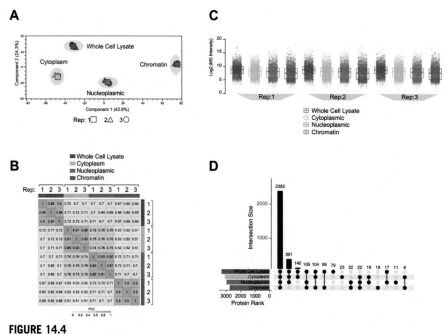

FIGURE 14.4

High-resolution mass spectrometry strategy enables in-depth proteome coverage.

2. LFQ values are extracted from the protein group table. In Perseus software, the LFQ values are transformed log2 and a protein is considered quantified only if it is detected in at least two out of three biological replicates.

3. Missing values imputation is carried out using a normal distribution strategy (width = 0.3; shift = 1.8), and a two-tailed t test applied with correction for multiple testing (Benjamini) [35].

(A) Principal Component Analysis (PCA) of the dataset. Each data point represents a single replicate ($n = 3$). Color subgroups represent each subcellular fractionation, including total lysate, cytoplasmic, nucleoplasmic, and chromatin, respectively. The shape of the points indicates different biological replicates. The oval gray color behind each replicate highlights a distinct separation between each subfraction sample.

(B) Pearson correlation coefficient showing the relationship between the subfraction proteome. The positive correlation coefficient is displayed in blue and reduced values are shown in white.

(C) Box plots of protein expression profile across all samples. Color subgroups represent each proteome subfraction.

(D) Overlap of proteome sets across each proteome subfraction, using an UpSet plot. 2382 proteins are commonly detected among all proteomes, while various intersample combinations are displayed in black.

At the bottom of the panel, a histogram shows the number of proteins detected and quantified in each subfraction proteome using our mass spectrometry strategy.

6. Advantages

Sample preparation is the only segment in the proteomic workflow that still largely relies on a series of manual handling and pipetting steps. The integration of automated processes alleviates many limitations that are associated with manual processing. Here the combination of the KingFisher system (protein purification) and the Evosep system (LC chromatography) reduces the time needed for sample preparation, enabling high throughput analysis by increasing sample reproducibility and the reliability of results. Depending on the experiment design, our strategy permits upscaling of the amount of sample measurement per day (up to 96) and enables high-throughput protein quantification, which can be beneficial in the clinical setting when studying proteins involved in different diseases and conditions.

Data-independent acquisition (DIA) methods have gained great popularity in bottom-up quantitative proteomics, as they overcome the routine data-dependent acquisition strategy (DDA). We conducted a successful quantitative proteomics strategy using a timsTOF Pro mass spectrometer based on the diaPASEF method. Data are analyzed using DIA-NN software, with the aim of increasing proteome coverage and protein quantification accuracy in a relatively short time frame.

7. Limitations

The cell prefractionation strategy enables reduced sample complexity that allows for the detection of the less abundant proteins within a cell.

Subcellular fractionation is often accompanied by substantial losses of subcellular components, and therefore scaling of protein abundance from subcellular fractions to whole cell or organ abundances has been very challenging. For example, some degree of protein mixing or relocalization during the fractionation process is possible. Nonetheless, the data interpretation is still very challenging, since only a few reliable protein markers that correspond to cell compartments are available. The accumulation of data, appropriately combined, increases the resolution and confidence of subcellular cell fractionation as a powerful tool for understanding the organization of the cell and promoting in-depth proteome coverage.

With the increasing number of high-throughput MS analyses being carried out, data needs to be efficiently stored and adequate software is necessary to promote real-time data analysis and interpretation.

We recommended powerful computer drive data processing to create more accurate and large-scale data analysis and obtain a confident biological interpretation of the results.

8. Safety considerations and standards

Despite vast technical improvements, MS-based proteomics is not yet routinely established in a clinical environment [6].

The lack of a reporting standard is a fundamental problem for LC—MS methodologies.

For example, the number of instrument configurations is nearly infinite and different analytical strategies can substantially affect the data interpretation.

Thus, creating durable and accurate analyses of mass spectrometry data across several analytical/clinical facilities is still extremely challenging.

We emphasized our approach as compatible and complementary with some strategies already present in clinical laboratories [36].

9. Alternative methods/procedures

A variety of technical options to address protein quantification use chemical labeling. Most commonly, stable isotopic labeling is performed by metabolic labeling using stable isotope labeling with amino acids in cell culture (SILAC) [37], chemically stable isotopic dimethylation [38], and isobaric labeling using isobaric tagging for relative and absolute quantification (iTRAQ) [39], or tandem mass tags (TMT) [40].

The great advantage of isobaric tagging is that samples can be labeled separately with different isotopes and then combined before injection into the mass spectrometer. The samples can therefore be co-analyzed, and the relative quantification occurs within a single experiment rather than between multiple runs. This intrinsically leads to higher reproducibility and avoids the missing value problem which is frequently a major limitation of the DDA label-free strategy [41].

A key benefit of isobaric labeling is increased analytical throughput. However, some limitations remain. One disadvantage is the reliable quantification of low-abundance proteins across multiple samples. Also, methods that use chemical labeling can be less cost effective.

10. Troubleshooting and optimization/notes

Note 1: To isolate the chromatin, increasing concentrations of NaCl are added. NaCl works by disrupting the charge-based interactions between DNA and proteins, and also the charge-based interactions between proteins [42,43].

Proteins weakly bound to the chromatin are eluted off at low salt concentration (50—150 mM), while strongly bound proteins are eluted off at high salt concentration (200—400 mM). Proteins that are high-salt insoluble will remain in the pellet after the high salt treatment. Use EDTA to elute these, to disrupt the native chromatin structure [44,45]. 600 mM of NaCl may affect the protein-to-protein interaction efficiency, therefore endogenous immunoprecipitation assay is not recommended.

Note 2: Centrifugation is performed at room temperature to avoid precipitation of urea.

Note 3: The iodoacetamide solution should be prepared fresh prior to each experiment.

Note 4: Cells required: 5–10 million per each subcellular compartment, 80%–90% confluent.

Note 5: For whole cell lysate protein extraction take a small aliquot of suspended cells, usually between 5% and 10% of the total cell pellet amount.

Note 6: Avoid foaming of samples in both homogenization (tissues) and sonication (cells or tissues) steps.

Note 7: To check whether the cell membranes have burst, take 10 μL of sample and mix with 10 μL of trypan blue. Look under a microscope and observe whether there are intact cells (trypan blue will not penetrate these, whereas it will stain the nuclei of the burst cells). If there are intact cells, add more Buffer A or else incubate for longer on ice.

Note 8: Gently break the cell membrane, taking care to not twist (could break nuclei) or make bubbles. Take 10 μL of the sample from the Dounce and mix with 10 μL of trypan blue to check whether all the cells are lysed (see Note 1).

Note 9: Nuclei (and some heavy mitochondria) will pellet at $500–1000 \times g$ (as will intact cells). To check if all the nuclei pelleted, take 10 μL of the supernatant and use the trypan blue method (see Note 1). If nuclei remain in the supernatant, spin at $700 \times g$ for a further 5–10 min and then check again. Nuclei form a white pellet while intact cells form a yellow/brownish pellet. If some intact cells are observed, remove the supernatant carefully (retain for cytoplasmic fraction if desired) and add 1 mL of fresh Buffer A. Incubate on ice for 10 min then check again.

Note 10: In the MNase fraction there may be proteins which are not related to the chromatin environment. These proteins usually belong to the nucleoplasm and the surrounded nuclear environment.

Note 11: Benzonase incubation time may be adjusted due to the amount of DNA/RNA present in the sample.

Note 12: Sidoli et al. described a fully integrated workflow for characterizing histone posttranslational modifications (PTMS) using mass spectrometry (MS) [29]. We recommend the histone purification workflow when each subcellular fractionation buffer also contains the following components, at final concentration: 5 M Sodium butyrate, $1\times$ of Deacetylase Inhibitors, and 200 mM AEBSF.

Note 13: Typically, samples should have a protein concentration of [Protein] = 1 mg/mL. Protein quantification should be performed to determine protein concentration of each sample. We recommend digesting between 20 and 25 μg of starting material.

Note 14: We recommend resuspending the beads in 100% pure ethanol, at the working concentration of 10 μg/μL.

Note 15: Usually, beads are clumped or aggregated at the bottom of the cone plate. Place the sample on the magnetic rack again, wait for the liquid to clear (approx. 30 s) and move in a new low binding tube.

11. Summary

In this chapter, we describe an MS-based clinical proteomic strategy for analyzing the proteome profile of a tumor cell. Using the automation platform, we achieved a significant depth of proteome coverage in a relatively short time frame. In total we measured 3371 proteins, across 12 samples in less than 3 h of MS data acquisition time. This compares well against the canonical MS acquisition time of 1 h per sample.

Our aim is to describe a mass spectrometry analytical strategy that can be applied to acquire valuable information across "multi-omic" clinical studies and contribute to taking comprehensive "snapshots" of the cellular proteome environment. The benefit of our approach is the development of a good reproducibility proteome resource used to detect early disease and consequently provide a patient-tailored therapy.

Acknowledgments

The research reported was supported by The Comprehensive Molecular Analytical Platform (CMAP) under The SFI Research Infrastructure Programme, reference 18/RI/5702.

Author contributions

G.O. Conceptualization, Investigation, Methodology, Visualization, Writing—original draft, Writing—review and editing.

K.W. Methodology, Writing—review and editing.

P.M. Review and editing.

References

[1] Aebersold R, Mann M. Mass-spectrometric exploration of proteome structure and function. Nature 2016;537(7620):347−55. https://doi.org/10.1038/nature19949. 537 (2016).

[2] Cravatt BF, Simon GM, Yates JR. The biological impact of mass-spectrometry-based proteomics. Nature 2007;450:991−1000. https://doi.org/10.1038/nature06525.

[3] Nilsson T, Mann M, Aebersold R, Yates JR, Bairoch A, Bergeron JJM. Mass spectrometry in high-throughput proteomics: ready for the big time. Nat Methods 2010;7:681−5. https://doi.org/10.1038/NMETH0910-681.

[4] Wysocki VH, Resing KA, Zhang Q, Cheng G. Mass spectrometry of peptides and proteins. Methods 2005;35:211. https://doi.org/10.1016/j.ymeth.2004.08.013.

[5] Zhang Y, Fonslow BR, Shan B, Baek MC, Yates JR. Protein analysis by shotgun/ bottom-up proteomics. Chem Rev 2013;113:2343. https://doi.org/10.1021/CR3003533.

[6] Krasny L, Huang PH. Data-independent acquisition mass spectrometry (DIA-MS) for proteomic applications in oncology. Mol Omi 2021;17:29–42. https://doi.org/10.1039/D0MO00072H.

[7] Slavov N. Increasing proteomics throughput. Nat Biotechnol 2021;39(7):809–10. https://doi.org/10.1038/s41587-021-00881-z.

[8] Leutert M, Rodríguez-Mias RA, Fukuda NK, Villén J. R2-P2 rapid-robotic phosphoproteomics enables multidimensional cell signaling studies. Mol Syst Biol 2019;15. https://doi.org/10.15252/MSB.20199021.

[9] Müller T, Kalxdorf M, Longuespée R, Kazdal DN, Stenzinger A, Krijgsveld J. Automated sample preparation with SP3 for low-input clinical proteomics. Mol Syst Biol 2020;16:e9111. https://doi.org/10.15252/MSB.20199111.

[10] Wojtkiewicz M, Berg Luecke L, Kelly MI, Gundry RL. Facile preparation of peptides for mass spectrometry analysis in bottom-up proteomics workflows. Curr Protoc 2021; 1. https://doi.org/10.1002/CPZ1.85.

[11] Liu X, Gygi SP, Paulo JA. A semiautomated paramagnetic bead-based platform for isobaric tag sample preparation. J Am Soc Mass Spectrom 2021;32:1519–29. https://doi.org/10.1021/JASMS.1C00077.

[12] Strasser L, Oliviero G, Jakes C, Zaborowska I, Floris P, Ribeiro da Silva M, et al. Detection and quantitation of host cell proteins in monoclonal antibody drug products using automated sample preparation and data-independent acquisition LC-MS/MS. J Pharm Anal 2021. https://doi.org/10.1016/J.JPHA.2021.05.002.

[13] Gaun A, Lewis Hardell KN, Olsson N, O'brien JJ, Gollapudi S, Smith M, et al. Automated 16-plex plasma proteomics with real-time search and ion mobility mass spectrometry enables large-scale profiling in naked mole-rats and mice. J Proteome Res 2021;20:1280–95. https://doi.org/10.1021/ACS.JPROTEOME.0C00681.

[14] Li KW, Gonzalez-Lozano MA, Koopmans F, Smit AB. Recent developments in data independent acquisition (DIA) mass spectrometry: application of quantitative analysis of the brain proteome. Front Mol Neurosci 2020;13:564446. https://doi.org/10.3389/FNMOL.2020.564446.

[15] Gillet LC, Navarro P, Tate S, Röst H, Selevsek N, Reiter L, et al. Targeted data extraction of the MS/MS spectra generated by data-independent acquisition: a new concept for consistent and accurate proteome analysis. Mol Cell Proteomics 2012;11. https://doi.org/10.1074/mcp.O111.016717. O111.016717.

[16] Ludwig C, Gillet L, Rosenberger G, Amon S, Collins BC, Aebersold R. Data-independent acquisition-based SWATH-MS for quantitative proteomics: a tutorial. Mol Syst Biol 2018;14. https://doi.org/10.15252/MSB.20178126.

[17] Barkovits K, Pacharra S, Pfeiffer K, Steinbach S, Eisenacher M, Marcus K, et al. Reproducibility, specificity and accuracy of relative quantification using spectral library-based data-independent acquisition. Mol Cell Proteomics 2020;19:181–97. https://doi.org/10.1074/MCP.RA119.001714.

[18] Muntel J, Gandhi T, Verbeke L, Bernhardt OM, Treiber T, Bruderer R, et al. Surpassing 10 000 identified and quantified proteins in a single run by optimizing current LC-MS instrumentation and data analysis strategy. Mol. Omi. 2019;15:348–60. https://doi.org/10.1039/C9MO00082H.

[19] Lou R, Tang P, Ding K, Li S, Tian C, Li Y, et al. Hybrid spectral library combining DIA-MS data and a targeted virtual library substantially deepens the proteome coverage. iScience 2020;23:100903. https://doi.org/10.1016/J.ISCI.2020.100903.

[20] Zhang F, Ge W, Ruan G, Cai X, Guo T. Data-independent acquisition mass spectrometry-based proteomics and software tools: a glimpse in 2020. Proteomics 2020;20:1900276. https://doi.org/10.1002/PMIC.201900276.

[21] Demichev V, Messner CB, Vernardis SI, Lilley KS, Ralser M. DIA-NN: neural networks and interference correction enable deep proteome coverage in high throughput. Nat Methods 2019;17(1):41–4. https://doi.org/10.1038/s41592-019-0638-x.

[22] Searle BC, Swearingen KE, Barnes CA, Schmidt T, Gessulat S, Küster B, et al. Generating high quality libraries for DIA MS with empirically corrected peptide predictions. Nat Commun 2020;11(1):1–10. https://doi.org/10.1038/s41467-020-15346-1.

[23] Gao M, Yang W, Li C, Chang Y, Liu Y, He Q, et al. Deep representation features from DreamDIAXMBD improve the analysis of data-independent acquisition proteomics. Commun Biol 2021;4(1):1–10. https://doi.org/10.1038/s42003-021-02726-6.

[24] Alberts B, Johnson A, Lewis J, Raff M, Roberts K, Walter P. Fractionation of cells. 2002. https://www.ncbi.nlm.nih.gov/books/NBK26936/. [Accessed 8 December 2021].

[25] Lee YH, Tan HT, Chung MCM. Subcellular fractionation methods and strategies for proteomics. Proteomics 2010;10:3935–56. https://doi.org/10.1002/PMIC.201000289.

[26] Masuda T, Sugiyama N, Tomita M, Ohtsuki S, Ishihama Y. Mass spectrometry-compatible subcellular fractionation for proteomics. J Proteome Res 2020;19:75–84. https://doi.org/10.1021/ACS.JPROTEOME.9B00347.

[27] Bache N, Geyer PE, Bekker-Jensen DB, Hoerning O, Falkenby L, Treit PV, et al. A novel LC system embeds analytes in pre-formed gradients for rapid, ultra-robust proteomics. Mol Cell Proteomics 2018;17:2284–96. https://doi.org/10.1074/MCP.TIR118.000853.

[28] Meier F, Brunner AD, Frank M, Ha A, Bludau I, Voytik E, et al. diaPASEF: parallel accumulation–serial fragmentation combined with data-independent acquisition. Nat Methods 2020;17(12):1229–36. https://doi.org/10.1038/s41592-020-00998-0. 17 (2020).

[29] Sidoli S, Bhanu NV, Karch KR, Wang X, Garcia BA. Complete workflow for analysis of histone post-translational modifications using bottom-up mass spectrometry: from histone extraction to data analysis. JoVE (Journal Vis Exp) 2016:e54112. https://doi.org/10.3791/54112. 2016.

[30] Rappsilber J, Mann M, Ishihama Y. Protocol for micro-purification, enrichment, pre-fractionation and storage of peptides for proteomics using StageTips. Nat Protoc 2007;2(8):1896–906. https://doi.org/10.1038/nprot.2007.261.

[31] Bekker-Jensen DB, Martínez-Val A, Steigerwald S, Rüther P, Fort KL, Arrey TN, et al. A compact quadrupole-orbitrap mass spectrometer with FAIMS interface improves proteome coverage in short LC gradients. Mol Cell Proteomics 2020;19:716–29. https://doi.org/10.1074/MCP.TIR119.001906.

[32] Meier F, Brunner A-D, Koch S, Koch H, Lubeck M, Krause M, et al. Online parallel accumulation–serial fragmentation (PASEF) with a novel trapped ion mobility mass spectrometer. Mol Cell Proteomics 2018;17:2534–45. https://doi.org/10.1074/mcp.TIR118.000900.

[33] Cox J, Hein MY, Luber CA, Paron I, Nagaraj N, Mann M. Accurate proteome-wide label-free quantification by delayed normalization and maximal peptide ratio extraction, termed MaxLFQ. Mol Cell Proteomics 2014;13:2513. https://doi.org/10.1074/MCP.M113.031591.

[34] Tyanova S, Temu T, Sinitcyn P, Carlson A, Hein MY, Geiger T, et al. The Perseus computational platform for comprehensive analysis of (prote)omics data. Nat Methods 2016. https://doi.org/10.1038/nmeth.3901.

[35] Oliviero G, Brien GL, Watson A, Streubel G, Jerman E, Andrews D, et al. Dynamic protein interactions of the polycomb repressive complex 2 during differentiation of pluripotent cells. Mol Cell Proteomics 2016. https://doi.org/10.1074/mcp.M116.062240. mcp.M116.062240.

[36] Vogeser M, Schuster C, Rockwood AL. A proposal to standardize the description of LC−MS-based measurement methods in laboratory medicine. Clin Mass Spectrom 2019;13:36. https://doi.org/10.1016/J.CLINMS.2019.04.003.

[37] Ong SE, Blagoev B, Kratchmarova I, Kristensen DB, Steen H, Pandey A, et al. Stable isotope labeling by amino acids in cell culture, SILAC, as a simple and accurate approach to expression proteomics. Mol Cell Proteomics 2002;1:376−86. https://doi.org/10.1074/MCP.M200025-MCP200.

[38] Hsu JL, Huang SY, Chow NH, Chen SH. Stable-isotope dimethyl labeling for quantitative proteomics. Anal Chem 2003;75:6843−52. https://doi.org/10.1021/AC0348625.

[39] Ross PL, Huang YN, Marchese JN, Williamson B, Parker K, Hattan S, et al. Multiplexed protein quantitation in *Saccharomyces cerevisiae* using amine-reactive isobaric tagging reagents. Mol Cell Proteomics 2004;3:1154−69. https://doi.org/10.1074/MCP.M400129-MCP200.

[40] Thompson A, Schäfer J, Kuhn K, Kienle S, Schwarz J, Schmidt G, et al. Tandem mass tags: a novel quantification strategy for comparative analysis of complex protein mixtures by MS/MS. Anal Chem 2003;75:1895−904. https://doi.org/10.1021/AC0262560.

[41] Pappireddi N, Martin L, Wühr M. A review on quantitative multiplexed proteomics. Chembiochem 2019;20:1210. https://doi.org/10.1002/CBIC.201800650.

[42] Sanders MM. Fractionation of nucleosomes by salt elution from micrococcal nuclease-digested nuclei. J Cell Biol 1978;79:97−109. https://doi.org/10.1083/JCB.79.1.97.

[43] Herrmann C, Avgousti DC, Weitzman MD. Differential salt fractionation of nuclei to analyze chromatin-associated proteins from cultured mammalian cells. Bio-Protocol 2017;7. https://doi.org/10.21769/BioProtoc.2175.

[44] Teves SS, Henikoff S. Salt fractionation of nucleosomes for genome-wide profiling. Methods Mol Biol 2012;833:421−32. https://doi.org/10.1007/978-1-61779-477-3_25.

[45] Oliviero G, Kovalchuk S, Rogowska-Wrzesinska A, Schwämmle V, Jensen ON. Distinct and diverse chromatin proteomes of ageing mouse organs reveal protein signatures that correlate with physiological functions. Elife 2022;11. https://doi.org/10.7554/ELIFE.73524.

Index

'Note: Page numbers followed by "f " indicate figures and "b" indicate boxes.'

219